THE NEW NORMAL

Pressures on Technical Communication Programs in the Age of Austerity

Edited by:
Denise Tillery
and
Ed Nagelhout

University of Nevada, Las Vegas

Baywood's Technical Communications Series
Series Editor: CHARLES H. SIDES

Routledge
Taylor & Francis Group
LONDON AND NEW YORK

First published 2015 by Baywood Publishing Company, Inc.

Published 2017 by Routledge
2 Park Square, Milton Park, Abingdon, Oxon OX14 4RN
711 Third Avenue, New York, NY 10017, USA

Routledge is an imprint of the Taylor & Francis Group, an informa business

Library of Congress Catalog Number:
ISBN 13: 978-0-89503-914-9 (hbk)
ISBN 13: 978-0-89503-915-6 (pbk)

Library of Congress Cataloging-in-Publication Data

The New Normal : pressures on technical communication programs in the age of austerity / Edited by:
Denise Tillery and Ed Nagelhout, University of Nevada, Las Vegas.
pages cm
Includes index.
ISBN 978-0-89503-914-9 (cloth : alk. paper) -- ISBN 978-0-89503-915-6 (pbk. : alk. paper) -- ISBN 978-0-89503-916-3 (epub) -- ISBN 978-0-89503-917-0 (epdf)
1. Communication of technical information--Study and teaching (Higher)--United States. 2. Education, Higher--Economic aspects. 3. Universities and colleges--United States--Cost control. I. Tillery, Denise, editor. II. Nagelhout, Ed, editor.
T10.63.A1N49 2015
607.1'173—dc23
2015029903

Table of Contents

INTRODUCTION

Austerity and Marginalized Academic Programs Why This Collection? Why Now?

Denise Tillery and Ed Nagelhout

As colleges and universities across the country continue to deal with regular decreases in state funding, technical communication programs, in particular, are being forced—as the theme of a recent CPTSC conference suggests—to "do more with less." As budget cuts become the new normal, the long-term health of technical communication depends on our ability to evolve and adapt to an array of internal, external, and technological pressures. This edited collection investigates how programs throughout the country are responding to challenges posed by the Great Recession and the new age of austerity, as well as the ways that universities and technical communication programs use increasingly limited resources to meet the challenges of increased student demand, online education, and the constant pressure to prepare our students appropriately for the ever-changing needs of the job market.

If the 1990s was the age of the dot-com boom, a late 20th century Gilded Age, we can call our current time the Age of Austerity. The term "austerity" is generally used to describe international financial policy. The concept, which entered the public consciousness through international news, describes a controversial set of financial policies that dramatically increase taxes and decrease government spending on social services in order to force countries such as Greece (or more recently, Cyprus) to start paying down overwhelming national debt. The United States federal government has not taken the drastic steps seen in Europe, although the 2013 "budget sequestration" is an unprecedented step in that direction. More importantly for higher education, these types of public spending cuts have been forced through state and local legislatures at an alarming rate. While this is not the forum for an extended analysis of the controversy surrounding policies implemented in the name of austerity, for our purposes in this collection, we use austerity as a shorthand term to describe the cumulative effects of lower levels of public funding for education, lower levels of grants and scholarships available to students to offset increased tuition, and the range of effects that contracting budgets create in academic contexts. Although the Great Recession may officially be over in the United States, this austere state of affairs has continued and has essentially become the "new normal."

This "new normal" is supposed to usher in long-term "sustainability," a concept most often associated with environmental causes, but one often borrowed in discussions of austerity measures; German Prime Minister Angela Merkel has been particularly fond of discussing measures in terms of "sustainable growth." Sustainability implies a state of dynamic balance, wherein any resources that are harvested from a system are replenished at the same rate. In economic terms, it suggests that countries should avoid the boom-and-bust cycle of the late 1990s or the mid-2000s, avoiding dramatic economic growth that is inevitably followed by a painful crash and striving instead for moderate, balanced growth.

The recession initiated by the financial crisis of 2007 has had serious impacts on higher education throughout the country. For example, a "survey of colleges conducted in December 2008 by *The Chronicle* and Moody's Investors Service found that nearly 50 percent of respondents had imposed some form of freeze on faculty hiring" (Shieh, 2013, p. 1). Even before the recession, however, many states had already pulled back on the percentage of their budgets allocated to higher education. The recession simply aggravated that trend, as states had less money to put toward anything. In March 2013, *The Atlantic* published a graph charting the shocking decline of state spending on higher education. Twenty-five states have cut spending on higher education at least 27.7%, all the way up to 50%, and 48 states have cut their higher education budgets extensively, the only exceptions to the trend being Wyoming and North Dakota (Weissman, 2013). As Weissman (2013) posits,

> Again, there are some people who might think these cuts are overdue. Others might simply argue that states, needing to balance their budgets, didn't have a choice. But I'd argue that these numbers are a vivid demonstration of why Washington's post-recession path has been so disastrous. Instead of taking advantage of historically low borrowing rates and aiding the states, Congress cut the lifeline once the first round of stimulus funding dried up. Graphs like these show us the consequences.

To address these kinds of budget shortfalls, many policymakers at the state and local levels have considered (and are still considering) a wide range of options. One brief from the American Association of State Colleges and Universities (AASCU) lists the 10 most important state policy issues for higher education (AASCU, 2012, pp. 2–5):

#1—State operating support for public higher education
#2—Productivity (the cost of providing a college education)
#3—Governance restructuring and regulatory reform
#4—College completion
#5—Performance-based funding
#6—Tuition policy
#7—State student grant program funding and reform
#8—College readiness
#9—Veterans education
#10—Immigration policy

While these are all important issues, they do take the focus away from "education." In a new age of austerity, the economics of higher education too often become the driving force behind policy initiatives, too often dictate the goals for what is possible, too often hamstring a long-term public asset in the name of short-term (and possibly short-sighted) fiscal responsibility, too often place the burden of fiscal responsibility on the most vulnerable.

In fact, one of the themes that emerges in media coverage of austerity is that it is the most vulnerable populations who are the most common targets. Cuts to welfare benefits (Campbell, 2013), pay cuts to workers (Itano, 2013), and cuts to education funding at all levels from pre-kindergarten to higher education are more and more common throughout Europe and the United States (Bryant, 2011; Weissman, 2013). Cuts in education are often focused particularly on programs that serve children with special needs or programs that are deemed insufficiently academic, such as music, art, and physical education, despite demonstrated benefits in improving academic performance (Bryant, 2011, pp. 12–13). This pattern of targeting vulnerable populations and programs also plays out in higher education settings. Smaller institutions, smaller programs within large institutions, and less protected populations, including both students and adjuncts and part-time instructors, are all at risk during times of austerity. This collection focuses on these points of vulnerability, asking not only how we should respond to austerity but also what we lose or give up when we respond in certain ways.

Many of us in technical communication and professional writing programs have experienced individual stresses in the form of pay cuts, increased teaching loads, and faculty positions going unfilled as colleagues retire or change jobs. The mantra of "doing more with less" is heard everywhere, and most of us are struggling individually with the challenge that universities are facing in the bigger picture: rising enrollments joined with falling revenues.

As institutions, colleges and universities are notoriously slow to adapt to change, while our social and economic landscapes are transforming rapidly. Academic programs in technical communication have some experience with the challenges of keeping up in a rapidly changing context; that has always been one of the major challenges in designing curricula and degree outcomes and objectives in our field.

Programs in technical communication can be particularly vulnerable when austerity measures hit universities because we often lack a comfortable fit in a home department, whether that department is English or Communications. If a program has its own department, it is often relatively new and lacks full institutional support. As Maylath, Grabill, and Gurak (2010) explain,

> A historical weakness of interdisciplinary programs is that they are not "owned" by powerful departments and are therefore vulnerable and weak. These programs can be particularly weak if departments charged with supporting them have it within their power to disown programs, whether by hook or by crook. (p. 277)

And in times of austerity, even supportive departments may not be able to protect weak or vulnerable programs. A frequent metaphor heard in budget-cutting times is an emphasis on the "low-hanging fruit" (e.g., Wassom, 2013). Weak or unprotected programs and small institutions make for very easy pickings.

On the other hand, technical communication as a discipline has always been strongly pragmatic, originating as it does in a service course for technical disciplines (Miller, 1979). Even before the recession, we were willing to explore practice-based classroom techniques such as service learning (Bowden & Scott, 2003) or new teaching methods, including online teaching (Cargile Cook & Grant-Davie, 2003; Tillery & Nagelhout, 2013). Our programs had already learned how to experiment, how to be flexible, and how to create arguments about our value, and those skills have served us well in the current period of dramatic transformations in university systems.

This collection, as a whole, attempts to answer, or at least address, questions such as, How have we been doing since the recession, and How should we plot the course for our future? Our introduction will conclude by tracing the rapid expansion and contraction of programs in technical communication from the 1980s to the present, discussing some lessons learned from this tendency to be tightly associated with a market in rapid flux. We will describe how our contributors respond to various challenges in their institutional settings, and finally, we will conclude with some unsolved problems, acknowledging the price that austerity has forced us to pay. We must be realistic about the choices we are making in order to determine the best path forward for our programs and our discipline.

AUSTERITY AND TECHNICAL COMMUNICATION: EXPANSION AND CONTRACTION OF PROGRAMS; LESSONS LEARNED?

As a discipline, we have mapped the terrain of our history quite thoroughly, thanks to the work of scholars like Kynell-Hunt and Savage (2003). From the beginning of engineering education associated with the Morrill Act of 1862, through the impact of the GI Bill, to the personal computer boom of the 1980s, we've told ourselves the story of our field's rapid growth and its search for a comfortable home in various academic departments. Table 1 offers data from two studies and recent views of the Society for Technical Communication's (STC) database of academic programs; the data reveal rapid growth in programs over a 27-year period, particularly in the late 1990s to early 2000s, and its consequent retraction since 2003.

The data in Table 1 is necessarily imperfect; the STC programs database leaves it up to the individual academic programs to participate. However, the table does reveal a pattern of rapid growth and sudden slowdown. The number of programs more than doubled from 1976 to 1981. The slower growth up to 1997 is surprising, considering those were the boom years in the growth of the computer industry, but slower growth could be attributed to poor participation in the STC program database. The sudden jump in the five years between 1997 and 2003 suggests better participation as well as a rush on the part of colleges and universities to participate in the dot-com boom of the 1990s. We might consider many of those programs late to the party, but, as discussed earlier, the lag can most likely be attributed to the notoriously slow pace of universities and colleges when it comes to acting on curriculum and program matters.

Sadly, soon after the largest increase in technical communication program creation, as indicated in our table (a growth of 350%), the Great Recession hit, causing the biggest increase in unemployment since the 1930s. Membership in the

Table 1. Technical Communication Programs in Society for Technical Communication Database

Code	Number of academic programs	Source
1976	19	CPTSC (cited by McDowell, 2003)
1981	56	CPTSC (cited by McDowell, 2003)
1997	65	Humphreys (1997) (STC Survey of academic programs)
2003	230	STC Database (cited by McDowell, 2003)
2012	207	STC Database (not including certificate programs)

STC dropped to around 7,000 after a high of almost 40,000. Job prospects for jobs in technical communication are not worse than average, but as always, practicing technical communicators often find themselves vulnerable to layoffs in companies in which they are not perceived as central to operations. These trends in program growth and employment do not mean that we should not be training more students in technical communication, but they do mean that we should be mindful not to market our programs with promises of easily available, high-paying jobs. We should also be mindful about growing our programs in responsible, sustainable ways.

A related difficulty that academic programs in technical communication face is the fast pace of change in the professional landscape. In 2004, Allen and Benninghoff reported on a survey of concepts taught in technical communication programs. Their top principles and topics include audience, rhetorical analysis, collaboration, document design, genre, ethics, visual rhetoric, user-centered design, information design, user awareness, and cultural differences. They also listed tools often taught, including Word, Photoshop, Dreamweaver, and Robohelp. Some of the principles, especially those based in traditional rhetorical theory, remain stable. However, some of them are new enough that many faculty whose degrees are more than 15 years old would not have learned them in their graduate training. Furthermore, some of the tools are already out of date. So even 10 years ago, faculty faced pressure associated with the need to learn new skills in order to teach them to students facing a rapidly changing marketplace. Some important topic areas, notably international and intercultural communication, have already been underrepresented in our curricula and faculty expertise.

These pressures have only increased as austerity measures hamper our ability to hire newly trained faculty, whether we are building programs or simply replacing retiring faculty. The pace of technical change has continued, or even accelerated, since the recession. We surveyed *Intercom* and *Technical Communication*, publications that are geared as much toward working professionals as academic researchers, and found that the following topics emerged between 2006 and 2013:

- Blogs, wikis
- Copyright/fair use/digital rights
- Tagging, metadata
- Podcasting
- Designing for mobile devices
- Social media
- User experience

Of course, new platforms such as blogs, wikis, and social media will necessarily depend on the same rhetorical skills of audience awareness, rhetorical abilities, and cultural sensitivity that we have always taught. But they also depend on increased skills in visual rhetoric, information design, and complex understandings of genre. And some concepts, including tagging and metadata, designing for mobile devices, and copyright and digital rights, require us to develop old skills in new ways or acquire entirely new fields of information. In a context of austerity, well-established programs with comprehensive curricula might keep programs current by redesigning courses to include new concepts and providing support for faculty training instead of hiring new faculty. For smaller programs, which may have already started with gaps in curricula and faculty expertise, austerity might mean falling even farther behind.

But curricula and faculty expertise are not the only areas of pressure on technical communication programs. External pressures, including a decreased trust in public institutions in general, lead to internal pressures, pressures applied to academic programs from within our institutions. While local pressures are always unique and dependent on local contexts, many programs, in academic terms, are feeling a range of pressures:

- Administrative pressures to increase students in degree programs
- The push to move courses into cheaper, online environments and compete with for-profit institutions, professional certification programs, and even other universities for students
- Increasing class sizes at all degree levels and on all platforms
- Emphasis on assessment requiring quantifiable measures of student success and consequent efforts to track and interpret that data
- Internal (within universities) funding formulas leading departments to see student credit hours as a zero sum game, leading to less collaboration and a tendency to decentralize writing instruction
- Increased teaching loads
- Well-established trends of universities to leave empty faculty positions unfilled and rely increasingly on part-time and adjunct faculty
- Increasing tenure standards that compel junior faculty to focus efforts on publish-or-perish
- Tenure committees that continue to privilege scholarly monographs and shy away from newer media, including online journals and digital research projects, which are becoming increasingly important in our field

In the face of such difficulties, technical communication programs have to be resourceful and imaginative, finding partners in new places and experimenting with new ways to reach students and meet their needs. Our contributors all demonstrate this willingness to experiment, to interrogate their institutions and local contexts, and formulate solutions to these problems. No solution is perfect, however, and we hope this collection will provoke discussions about what we have lost in addition to what we have gained as the age of austerity takes hold.

SOLUTIONS AND RESPONSES FROM OUR CONTRIBUTORS

Our 20 co-contributors hold a wide variety of academic positions: 8 assistant professors, 1 visiting assistant professor, 9 associate professors, and 2 full professors, with both faculty and administrative appointments. Similarly, our contributors are located in a range of institutions, from small undergraduate to master's-granting to high research institutions. Interestingly enough, only one contributor works at a flagship institution. We believe this reflects the distribution of, especially, undergraduate programs in technical communication. And while we all typically receive our graduate training at large (generally public) research institutions, we are highly likely to spend our academic careers in smaller colleges and universities that don't much resemble where we were trained. Most undergraduate students who are working toward a bachelor's in technical communication, or who take technical communication courses as part of another degree, can be found in these smaller (and often less expensive) institutions as well. Because this volume focuses on the impacts of austerity measures on vulnerable populations, it is fitting that most of our contributors are well situated to address the ways academic programs can respond within the local contexts of smaller, less well-funded institutions.

We've divided the chapters in this collection into three sections: Building Connections and Growing Sustainably; New Teaching Models: Adapting Technologies Strategically; and External Challenges and Opportunities.

Section 1, "Building Connections and Growing Sustainably," focuses on a common metaphor we found in responses to austerity: the importance of sustainable growth. This section addresses the ways that technical communication programs create a state of dynamic balance, as described earlier, in which any resources that are harvested from a system are replenished at the same rate. In terms of academic program administration, sustainability implies that we should not outgrow our local contexts. In boom years, when funding is plentiful and growth is encouraged, we should grow mindfully. For example, we should not market online classes before we have trained instructors to staff them, or offer multiple degrees before we have faculty expertise to make them meaningful. Sustainability also means that degree programs should serve students and communities, not just academic programs. That is, we should refrain from building undergraduate degree programs in an effort to chase student credit hours when there are few jobs available; similarly, we should avoid creating or expanding PhD programs simply because an academic institution desires the prestige of a graduate program.

This section begins with Barry Maid suggesting that although those of us in the field of technical communication often feel that our strong career emphasis makes us less vulnerable to threats from budget cuts, in reality, cuts that have been made since the start of the Great Recession have less to do with money and more to do with philosophy or vision. His chapter shows how and why technical communication programs need to make the program important to the institution by connecting with both internal and external constituencies. Having a clear understanding of their own institutional relevance as well as having strong connections can give technical communication faculty an advantage when times get tough.

Two chapters focus explicitly on the idea of sustainable growth. Amanda Bemer and Teresa Henning argue for a model of growth based in the concept of "ecopreneurship," a term that merges economic opportunity with responsible management of natural resources. This model views the program, the university, and the local context as an ecosystem, a dynamic set of relations that should be developed and nourished as a whole. In their view, a healthy program contributes to the system's overall health rather than competing for increasingly scarce resources in a battle that no one can ultimately win.

Colleen A. Reilly approaches austerity-related challenges by turning to the concept of systems thinking. She focuses on how her program has fostered interconnectedness, and addresses the structural and cultural barriers faculty face when they seek interdisciplinary connections within their own institutions.

Finally, while not directly impacted by cuts to state higher education allocations, writing programs housed at private colleges and universities have, like their state-supported brethren, felt the sting of austerity. Madeline Yonker and Michael J. Zerbe focus on three ways that austerity has impacted the undergraduate professional writing major and minor at York College of Pennsylvania. First, they trace the decline of enrollment in the professional writing major, discussing some of the austerity-induced pressures on enrollment at this institution. Second, they explore the impact of delaying technology (both hardware and software) upgrades on faculty and students in the professional writing program. Third, they investigate impediments to faculty-industry collaboration that have resulted from austerity.

Section 2, "New Teaching Models: Adapting Technologies Strategically," addresses the advantages as well as the challenges posed by the drive to incorporate new technologies into technical communication programs. This section offers a needed corrective to the idea that technologies are a panacea for economic challenges and budget cuts. While technologies like online courses offer accessibility to students with busy lives, and offer the opportunity for students (and instructors) to develop increasingly important information design and literacy skills, online education is neither cheap nor easy. Technical communication programs should resist the push to define such technologies as the solution to austerity problems.

This section begins with Julia Romberger and Shelley Rodrigo explaining and modeling a learning-to-learn approach as a rhetorical process that both develops scrappy students and better prepares students for whatever workplace environment graduates find themselves in. The approach leaves students better positioned to understand the need for lifelong learning because they have developed the ability to identify the networks of people, places, and things that allow them to access, learn, and use technologies when needed. Instead of graduating with a laundry list of applications they are "proficient" with, their students graduate with the skills and resources to adapt to the constantly changing technological landscape.

Tiffany Bourelle discusses strategies for adapting service learning pedagogy to the online environment. Service learning has long been an important way technical communication programs help students prepare for the job market; students learn pragmatic aspects of workplace writing that are impossible to reproduce in the classroom and at the same time, they make connections and gain practice writing for specific audiences. But in the transition to online classes, service learning projects can prove difficult to implement. Bourelle shows how service-Elearning occurs in her context, offering a model for other online programs to develop these opportunities for their own students.

Dirk Remley presents a case study of the efforts involved in developing an effective online course model that meets the needs of many stakeholders. As administrators constantly seek to increase efficiency, program directors are under pressure to make decisions about curricula and instructional models that may be met with resistance. Remley's case shows how important it is for program administrators to balance competing stakeholder interests and address frictions caused by departmental or institutional responses to austerity measures.

Our own chapter, written with Julie Staggers, focuses on the working conditions of adjuncts and part-time instructors as an important aspect of how we are responding to austerity in our own context. Our institution has always relied heavily on part-time instructors to cover service classes. As most of our adjuncts lack training in rhetoric, composition, or professional writing, we see faculty development as an important way to improve working conditions for adjunct instructors. There is little we can do to alleviate meager salaries, inadequate healthcare benefits, or lack of career stability for these hard-working teachers. We can, however, provide consistent faculty development to increase their skills, make them more efficient and effective inside and outside the classroom, and ensure they have a manageable work-life balance.

Finally, Section 3, "External Challenges and Opportunities," reports on innovative responses to a range of challenges posed by legislative mandates, new workplace realities, and changing student demographics. Many of the external challenges described in these chapters are brought about by highly specific local conditions, but the lessons learned from the responses to these challenges contribute to our ongoing conversations about how to define our discipline and how our discipline-specific skills will help us take an active part in shaping our own future. This section presents a diverse set of strategies for addressing external challenges, analyzing local contexts and problems, identifying opportunities, and changing in response to these conditions.

This section begins with Timothy Giles, Angela Crow, and Janice Walker describing a challenge to their program based in their university's response to budget cuts: outsourcing many information technology services to Google, adoptions that appear to be an emerging trend. The authors enumerate problems regarding intellectual property, professionalizing students, and privacy, which arise when students and faculty alike are forced to turn to Google for Web-server space, email services, and other applications formerly managed by the university's information technology support. Many programs are facing these problems, and Giles, Crow, and Walker articulate the important issues that we may overlook in our efforts to meet the daily needs of our programs.

Lynn O. Ludwig's context is shaped by a new state-mandated initiative to increase the number of healthcare providers in order to meet the needs of the state's population. Instead of seeing this initiative as a threat to her department's liberal arts mission, Ludwig took advantage of the opportunity for growth by developing an innovative program and identifying strategic partners to keep demand for the program high. She offers a model of how a small and institutionally disadvantaged program might respond to the needs of a local context and shape itself in innovative and sustainable ways.

Finally, Corinne Renguette, Marjorie Rush Hovde, and Wanda L. Worley describe the challenges of developing a new bachelor's degree in technical communication in the face of austerity. They confront the typical challenges austerity presents for many technical communication programs—a field that changes rapidly, due in part to the rapid changes in technology; institutional demands to add more online, hybrid, and untraditional delivery methods; a lack of resources; and a small number of full-time technical communication faculty members supplemented by part-time instructors. However, they also describe challenges brought on by new educational mandates from the Indiana State Legislature, as well as those created by their unique context: an institutional structure that combines two major state universities on one campus and their location in a school of engineering and technology.

UNSOLVED PROBLEMS AND PRICES PAID: WHAT DOES AUSTERITY FORCE US TO GIVE UP?

As faculty in a program that was dramatically threatened by state budget cuts several years ago (receiving, in fact, a layoff notice by email from our Dean), we are acutely aware that the sense of danger looming at the height of the recession has never entirely disappeared. We find it difficult to plan for more than a year in advance, because we cannot count on current faculty being around nor can we count on the ability to replace faculty who might leave for greener pastures. We understand better than most academics that tenure is not the guarantee of stability that it used to be—perhaps it never really was—and that we cannot continue to do business in the ways that we always have, confident that tenure lines will be added when student numbers grow, and assured that courses that have reliably full rosters every year will be offered in the same ways, at the same times, and to the same population as always. Our conditions have changed dramatically over the last 5 years, and we must change our responses to them.

When we sent out our open call for submissions, we did not anticipate that our responses would be primarily from smaller institutions, but upon reflection, we realized that austerity measures have hit smaller institutions much harder than big flagship institutions, which have the ability to find resources in other places. Our collection should provide valuable information and strategies for colleagues at those smaller programs, as well as for the larger institutions that are primarily responsible for training graduate students who will in many cases find jobs at these smaller, less well-supported programs.

This focus on smaller programs has left some unanswered questions, however, which we want to address here in the hopes that later publications will take them up and provide some imaginative solutions. One question that we did not cover in our collection is how we as faculty in technical communication can continue to develop our research interests. Many of us have experienced increasing teaching loads and decreasing wages, leaving less time to develop our own research agendas. Research is costly in many ways, although most of us do not need expensive equipment. We do need the time to develop questions, cultivate connections to research sites, analyze data, and write up findings. One way to gain time and resources is, of course, to acquire external funding. But developing and writing grants takes time, and many smaller programs are located in institutions that lack competitive resources. Austerity is also increasing pressure on grant-funding organizations as well, meaning that more and more individuals and groups are applying for a dwindling pool of funding. We believe that as researchers, we will need to look to local needs and local sources of data to pursue research in our field.

A second question involves the training of graduate students in our discipline. Our earlier research did not distinguish between different levels of degrees, so the number of new programs includes PhD programs as well as master's and bachelor's degrees. It is fair to say, however, that a number of new PhD programs in technical communication have been developed over the last 15 years. We hope that those programs are being developed in sustainable ways and that we are not simply producing PhDs for an academic marketplace that may not exist in the same form over the next several years. We hope to see conversations about how PhD training should develop in response to changes in academic and workplace structures, and to see less of an emphasis on training (cloning) PhD students to take the kind of high-prestige jobs their advisors have or perceive as desirable.

We hope to see shifts in what counts as success by many measures: an emphasis on quality rather than quantity of research publications; a focus on quality of life for students, faculty, and staff rather than measures of productivity, decontextualized grades or assessment data and other impersonal measures;

and a trust that academic institutions can be living, organic systems that evolve and grow in response to local contexts, rather than moribund, archaic institutions that fear and resist change.

REFERENCES

Allen, N., & Benninghoff, S. T. (2004). TPC program snapshots: Developing curricula and addressing challenges. *Technical Communication Quarterly, 13*(2), 157–185.

AASCU State Relations and Policy Analysis Team. (2012, January). Top 10 higher education state policy issues for 2012. *American Association of State Colleges and Universities.* Retrieved from http://www.aascu.org/uploadedFiles/AASCU/Content/Root/PolicyAndAdvocacy/PolicyPublications/Policy_Matters/Top_Ten_State_Policy_Issues_2012.pdf

Bowdon, M., & Scott, J. B. (2003). *Service-learning in technical and professional communication.* New York, NY: Longman.

Bryant, J. (2011, October). Starving America's public schools: How budget cuts and policy mandates are hurting our nation's students. *Campaign for America's Future, 13.* Retrieved from http://www.ourfuture.org/files/documents/starving-schools-report.pdf

Campbell, C. (2013, April 8). British politician claims he can live on $80 per week; Petition urging him to try hits 450,000. *Time.* Retrieved from http://newsfeed.time.com/2013/04/08/british-politician-claims-he-can-live-on-80-per-week-petition-ur/

Cargile Cook, K., & Grant-Davie, K. (Eds.). (2003). *Online education: Global questions, local answers.* Amityville, NY: Baywood.

Humphreys, D. (1997). *Education in scientific and technical communication: Academic programs that work* (M. Keene, Ed.). Arlington, VA: STC.

Itano, N. (2010, March 11). Greek austerity measures spark rising protests. *Time.* Retrieved from http://content.time.com/time/world/article/0,8599,1971491,00.html

Kynell-Hunt, T., & Savage, G. J. (Eds.). (2003). *Power and legitimacy in technical communication: The historical and contemporary struggle for professional status.* Amityville, NY: Baywood.

Maylath, B., Grabill, J., & Gurak, L. (2010). Intellectual fit and programmatic power: Organizational profiles of four professional/technical/scientific communication programs. *Technical Communication Quarterly, 19*(3), 262–280.

McDowell, E. E. (2003). Tracing the history of TC from 1850–2000, plus a series of survey studies. *ERIC Resources.* Retrieved from http://eric.ed.gov/?id=ED478117

Miller, C. R. (1979). A humanistic rationale for technical writing. *College English, 40*(6), 610–617.

Shieh, D. (2009, April 17). It's not just about the money: Faculty morale has been hurt by colleges' austerity measures, even if professors still have jobs. *The Chronicle of Higher Education.* Retrieved from http://chronicle.com/article/It-s-Not-Just-About-the-Money/34012

Tillery, D., & Nagelhout, E. (2013). Theoretically grounded, practically enacted, and well behind the cutting edge: Writing course development within the constraints of a campus-wide course management system. In K. Cargile Cook & K. Grant-Davie (Eds.), *Online education 2.0: Evolving, adapting, and reinventing online technical communication* (pp. 25–44). Amityville, NY: Baywood.

Wassom, R. (2013, March 14). Low hanging fruit should be picked from defense budget. *The Hill*. Retrieved from http://thehill.com/blogs/congress-blog/economy-a-budget/288171-low-hanging-fruit-should-be-picked-from-defense-budget

Weissmann, J. (2013, March 20). A truly devastating graph on state higher education spending: Some states have slashed per-student spending by as much as half. *The Atlantic*. Retrieved from http://www.theatlantic.com/business/archive/2013/03/a-truly-devastating-graph-on-state-higher-education-spending/274199/

SECTION ONE

Building Connections and Growing Sustainably

CHAPTER 1

Keeping the Target Off Our Backs: How to Build a Sustainable Technical Communication Program in Times of Austerity

Barry Maid

Those of us in the field of applied writing (a broad term I often use to encompass technical communication, professional writing, and really any kind of workplace writing) often feel that because what we teach is easily seen as connecting students to careers, we are more impervious to much of the budget-slashing frenzy that has gripped higher education for more than half a decade. The reality is that nothing is further from the truth.

I would like to suggest that though state portions of higher education budgets have been cut significantly, most institutions have found alternatives (most notably significantly higher tuitions) to compensate for the loss. I'd also like to suggest that cuts that have been made since the start of the Great Recession have less to do with money and more to do with philosophy or vision.

We can see one aspect of this change in vision that triumphs over economy in the phenomenon of "administrative bloat." There is no doubt that institutions of higher education suffered from economic stress as a result of the Great Recession and had to find various ways to survive. It's not unusual that when some of those cost-cutting measures affected class instruction because of a reduction in the number of faculty (both full- and part-time), that people like Bob Samuels, an officer in a faculty union, would note in his 2009 *CCC* piece that the same belt-tightening was not being experienced by those in administrative ranks (pp. A8–A10). However, Samuels is not alone in his perspective. We also find a report from the Goldwater Institute (at the other end of the political spectrum) that comes to the same conclusion—that administrative spending has been increased, not cut (Greene, Kisida, & Mills, 2010, p. A15). While this is just one example (increase of administrative bloat), my sense is, if perspectives as diverse as the Goldwater Institute and a labor union officer agree, that something is going on. We, and they, may disagree on the exigency for the changes. However, since they both document that more, not less, money is being spent in some areas (in this case administrative positions and salaries), then I think we can see that money is not the driving force in the "New Austerity." If this is the case, and I believe it is, then applied writing programs need to make themselves so important to their institutions that they can't become targets.

In this chapter, I propose to show how and why applied writing programs are important to institutions. I also intend to show how applied writing programs can connect with both internal and external constituencies. Having a clear understanding of their own institutional relevance as well as having strong connections can give applied writing faculty an advantage when times get tough. I will stress three areas of opportunity that, if followed, will help applied writing programs have a better chance of survival in the new "austere" world. They are

- Be relevant
- Be flexible
- Be connected

THE PROBLEM FOR APPLIED WRITING

Higher education in the United States may finally be changing. Many of us who have spent most of our adult lives working in higher education have known that change was necessary. We have often advocated for change within our own institutional contexts. Most often we've been met by resistance on the part of colleagues and administrators. The overwhelming mantra was, "things move slowly here." Perhaps moving slowly is understandable for an institution with roots in the Middle Ages. Yet the reality is that the rest of the world no longer moves at the speed it did in the 14th century. What may be even more important, however, is that whether we like it or not, institutions of higher education are no longer separate from the rest of society. They are completely intertwined with society. To ignore that fact, considering the current "austerity crisis," is no longer simply naïve—it's now suicidal. While an academic program such as technical writing is applied by its very nature and must connect to the outside world, this is no longer enough. Programs must actively engage.

Ed White has often mentioned his own "Law of College Assessment" when speaking of the need for good writing program assessment, to "Do assessment thyself or it shall be done unto thee" (1999). White's dictum holds for institutional change as well. Unfortunately, most institutions of higher education have chosen not to change—or to change only in minimal, superficial ways. As a result, external forces have begun to make significant changes on the American academy. This is not the place to make moral or ideological arguments. Rather, I'd prefer to simply identify two major forces working to change higher education. I do so merely to be informative and so that we may better understand the social, cultural, and economic landscape that surrounds change in higher education. The first force is those who have a political agenda to change higher education. It is not a monolithic force. Indeed, people from across the political spectrum may share the idea to change higher education—though their visions for change may differ. We can see how some of this plays out with the No Child Left Behind (NCLB) legislation which governs K–12. NCLB was promoted by those on both the left and the right, though clearly those of different political views had very different visions of the role of public education. The second major force exerting itself to help change higher education is those corporations that see higher education as a good place to make lots of money. Again, I think it would be wrong to see this force as monolithic. Companies who sell learning management systems (LMS), such as Blackboard, are very different from for-profit institutions like the University of Phoenix. Still, both of them and their competitors have become a very real part of the greater landscape of higher education.

So what does all of this have to do with an academic program in applied writing? My sense is a tremendous amount. Applied writing programs tend to be small—at least when compared to programs like psychology and communication. They are not part of the traditional liberal arts and science academy. However, they are often housed within traditional humanities units that have a historical disdain for anything applied. They also do not share the external community cachet of more traditional applied programs such as those in business and engineering. What that means is that when times get tough or need to get shaken up or both, applied writing programs tend to emerge as natural, easy targets. Without size, historical credibility, or outside constituencies to resist, administrators can easily reallocate resources from applied writing to anywhere else. What this means is that applied writing programs must not assume that being applied and having workplace connections will protect them. Applied writing programs must be proactive in creating a local environment in which the program is seen to be so integral to the institutional mission and the institution's stakeholders that it is unthinkable to cut.

In the past, when academic reorganization needed to go through multiple committees and a faculty senate, programs that found themselves to be potential targets had time to garner support. Those programs that were, indeed, underperforming with few students and fewer graduates would be questioned. Those programs that had sufficient enrollment and numbers of graduates had time to make their case. However, administrators have discovered that using "austerity" as an excuse, they can now make changes without going through an involved and lengthy process.

I offer up the case of my own program at Arizona State University as an example. When I was first hired to create a technical communication program at ASU, everyone in the administration from Dean to Provost talked of it as a "niche program." No one expected it to be huge. At ASU some majors have thousands of students. Still, a niche program is a good thing. In order to be a successful niche program we simply needed to define who we were, who our market was, and then serve that targeted market. We had done that. In fact, though I have related the story elsewhere (Maid & D'Angelo, 2013), it pays to repeat one part of our situation here. We were small, but growing; we could have grown quicker but were artificially constrained by the administration. I expect the two most significant administrative constraints to growth were geographic location and the inability to hire sufficient faculty.

I was hired in 2000 to create what was originally called the program in Multimedia Writing and Technical Communication that is located on (what is now called) the Polytechnic campus in southeast Mesa—23 miles from the Tempe campus and 34 miles from downtown Phoenix. While those distances are not great in terms of Western cities, they prove to be significant when students who work full-time have to deal with rush hour traffic to get to class on time. Since most of our students are nontraditional (average age of undergraduates was 33.6 in 2009), location was an obstacle to growth. I faced the location challenge by moving the program online. Our first online classes were taught in 2002. By 2004, all classes were available online; by 2009, we were entirely online.

The second artificial constraint had to do with our inability to hire enough faculty. Once we started offering online classes, it was clear that I could fill many more sections of some classes than I was offering. However, I was given a specific dollar figure that I could use to hire part-time faculty. When I went to my dean to request money for more sections, I was simply told there was no money. I knew better. I knew full sections taught by part-timers made money for the institution. However, that wasn't the way ASU, or at least my dean's view of ASU, worked. In the fall of 2008, I started working with a new dean who came from outside of ASU. He basically told me to offer as many sections of any class that would fill and for which I could find a qualified instructor. It looked as though we were going to grow faster. Then ASU hired a new provost, and instead of "niche programs" (which were good), she started talking about "boutique programs" (which were extravagant). Clearly, language makes a difference.

We can see a bit how this different view of our program nuanced by language affected the ensuing reorganization. The larger applied programs in our college (nutrition, exercise and wellness, and applied psychology) were moved to other colleges. Technical communication was combined with a small English and smaller history program whose main course offerings were to provide general studies courses to students at the Poly campus. The reality is, I suspect, the administration didn't know where to put us. If we had been a "niche," they would have found a way for our "niche" to thrive. However, since we were merely a "boutique"—possibly just a frill—we were placed with units that offered general studies humanities courses. Where frills are located can be an afterthought.

Overnight, we went from being a small, sustainable, and growing program to one that was seen as an expensive frill. The reasons this could happen are easily understood from what I've suggested above. We were small compared to other ASU programs. I suspect that any program with fewer than several hundred majors is seen as small by ASU. We are not a traditional academic discipline. Even in times of "austerity," it's more difficult to attack the traditional academic core than it is to go after applied programs. Even when administrative cuts are made unilaterally, it's much easier to get most faculty to go along when the traditional academy is kept intact. Finally, for a variety of reasons, we hadn't done our work in making strong external connections. The result was that when ASU decided to be "more efficient" since we had budget problems, we were reorganized from an independent unit with its own budget to a part of an amorphous unit called the Faculty of Interdisciplinary Humanities and Social Sciences. This unit is overseen by a Faculty Head, but because it is a "faculty" and not a department or a program, it has no budget. The Dean makes all budgetary decisions.

I tell this anecdote for several reasons. While ASU may be more extreme than other institutions, it clearly used the excuse of "austerity" in order to justify a restructuring that had nothing to do with efficiency or cost cutting. Rather, the changes reflect a philosophical change in how an institution of higher education should operate. What this means is that just being able to show that a program does not cost much money will not work. It's also possible that showing a program may produce revenue will be insufficient. In order to survive, a program must create a culture within the institution that it is too important to cut. It is rare that applied writing programs have done that.

TOO IMPORTANT TO CUT

In a world where maintaining an important niche can shift instantly to being a frivolous boutique, relevance—especially institutional and stakeholder relevance—becomes paramount. The reality for applied writing programs is, I think, that they seem, to the faculty within the programs, so obviously connected and relevant that they fail to make those connections known to others. Yet better public relations is only a part of the issue. Applied writing programs don't become connected or relevant just because the field assumes they are. Each program must cement its own relevance within its own academic and broader community. They must create tangible and acknowledged relationships with other academic units and with external stakeholders.

Too often the obvious is unspoken because everyone assumes it is known. Part of my agenda here is to identify and name what too many of us assume is obvious. The reality is that it's not always obvious. So first and foremost, all faculty in applied writing programs need to acknowledge that those programs exist for our students. Almost every student who goes through an applied writing program with the intention of getting a degree intends to use that degree to help succeed in the workplace. On some level, that may be true of any undergraduate degree. Still, students who major in applied writing are more likely to expect a direct academic/workplace connection than are students who major in philosophy, history, or even English. In reality, even students who major in psychology understand that their undergraduate work may only slightly relate to their eventual career. That does not mean that applied writing programs are or should be vocational. Vocational programs train students to perform a specific job usually entailing a limited number of tasks. If an applied writing program were vocational, it might train students to create and update user manuals from a predetermined template and nothing more. That's not to say that some applied writing graduates might not, at some time, find themselves creating or updating a manual from a template; however, their careers will most likely involve more complex writing tasks. In fact, applied writing graduates are more likely to be the ones who understand how to create an effective template rather than just filling in predetermined chunks.

So then, if we acknowledge that student needs are primary, we need to understand our students. We need to understand who they are, what they do, what they intend to get out of the program and a variety of other details of how they currently connect to the program and how they might continue to connect to the program even after they graduate. But, this should always be defined locally and contextually. Understanding student needs helps us to analyze whether the program we are attempting to deliver does, indeed, meet those student needs best. If we understand that the new "austerity" is a philosophical and ideological agenda and not about money, then we begin to realize that making a case for relevance must be done in current, not traditional, terms. In fact, it may even be possible to enhance your program. I'm a firm believer that sometimes the best time to expand is when everyone else is contracting.

WHO ARE YOUR STUDENTS?

One of the biggest myths I've encountered in my more than 40 years in higher education is that undergraduate students are 18–22 years old, live on campus, don't work, and come from families that support them. One of the consequences of that myth is that students should be able to declare a major from the moment they walk onto campus, if not before. It's a general cultural myth that most in higher education have bought into, even though they may see a very different reality every day. My experience is that very few applied writing majors fit that description. Having students who don't fit the mythical mold potentially causes problems. You can't and shouldn't change who your students are, but you should be able to understand who they are and proactively promote them before their difference becomes a liability.

In over 30 years of working with applied writing students, both undergraduate and graduate students, only once have I talked to a 17-year-old high school senior who said he was interested in majoring in technical communication. He knew about the program because several years earlier his uncle had graduated from the master's program. I expect he was one of the only high school seniors who even knew what technical communication was. Is it possible to recruit high school seniors to applied writing programs? I expect so; though I wonder if it's worth the time, effort, and expense. There are more than enough students who want to change their majors, transfer in from a 2-year college, or who are attracted to an applied writing major because of current or previous workplace experience. The problem is that those demographics don't fit the myth. More importantly, many institutions are less willing to put any resources into programs that attract those populations because those populations don't show up in federal graduation rates.

When understanding who your students are, it's important to also understand how different demographics may be treated by your institution as a result of reporting guidelines. For example, retention rates usually apply only to students who matriculate at an institution. Graduation rates definitely only count what are called "native students," that is, students who take their very first college course at an institution. A student who transfers any credits in (including dual-credit courses earned in high school) does not count toward a school's graduation rate. That means that returning students, who usually transfer in some credits, and community college students don't show up in the graduation rate—unless there's a returning student who happened to matriculate initially at your institution. The problem with those students is that unless they graduate within 6 years of graduation, even if they return and graduate, the time to graduation will hurt your institution's graduation rate. This doesn't mean that institutions don't want these students. They do. After all, all students pay tuition. However, institutions are more likely be put resources into programs that make them look good on national reports. Since it is the case that applied writing programs are unlikely to have students who make institutions look good on national reports, they need to become institutionally important in some of the other ways I will suggest. Again, the key is that traditional institutional values may no longer carry weight. You need to understand the current institutional values—even if those values make you a bit uncomfortable. Remember, understanding values doesn't mean embracing them.

WHAT IS YOUR PROGRAM LIKE?

This section assumes you have a program and are not just delivering one or several service courses. There's nothing wrong, indeed much is right, about service courses. However, the issue of maintaining service courses is a slightly different matter, and one that I'll address later. Applied writing programs are varied. Some are certificate programs, some are minors, some are tracks in majors, some are specific majors, some are master's programs, and some are doctoral programs. At some institutions, there are multiple programs. Usually, it's more difficult, though not impossible, to eliminate a program that offers both graduate and undergraduate degrees.

While PhD programs may be the most prestigious, especially within the academy, except for a handful of institutions, the most important degree to offer is an undergraduate degree. Alumni most often identify with the program or department from which they received their undergraduate degree. Regardless of whether your program is a designated set of courses or a degree, it's crucial that your curriculum is relevant and that others—students, administrators, alumni, and community stakeholders—know it. Not being relevant with any of those groups can cause problems. Just a few weeks ago, I had a phone conversation with a Dean of Arts and Sciences at a metropolitan university (not one where I have taught) in another part of the country. She wanted information about independent writing programs. As we talked, knowing that I had created the technical communication program at ASU, she asked me if I knew about the technical communication certificate offered by her English Department. Having reviewed all the writing curriculum prior to our call, I acknowledged that I was aware of it and had looked at the curriculum. I also acknowledged that I was acquainted with the faculty in the program. She asked what I thought of the program. I responded honestly that I thought it was a very good "early 2000s curriculum." She laughed and said she knew that was the case. (I think it's important to note that this dean's academic area is not writing.) She asked me what I thought it might take to update the curriculum. My response was to ask her if she knew why it hadn't been updated. Was it that faculty were too set in their ways? Was it that the curriculum was embedded in an English Department whose internal politics might make curricular revision difficult? Were there other external issues, either within the university or with external stakeholders, that might work against making the curriculum more current? She admitted she wasn't sure. The real point, however, is that this is an example of a dean who understands that there is value to a technical communication program but also understands that the program her institution is offering is dated. While I have no indication that this program is presently a target, my sense is that it's a program the current dean would like her college to offer. However, given the environment of "austerity" that faces all of us, unless the faculty updates the program, or it has some external champions, that program may be at risk. My current perspective is that we are all potential targets unless we actively engage in some of the strategies that I am proposing or other strategies that will obtain similar results.

Flexibility

When things shift, no matter whether those things might be geological plates, paradigms, or administrative ideologies, the ability to be flexible and adapt quickly is crucial for survival. Flexibility is a key component to survival in "austere times"—especially for academic programs. There are multiple ways technical communication programs can be flexible that will help them survive. Some of the areas in which I think Applied Writing programs might be most flexible are with curriculum, methods of delivery, students, and faculty.

I've just related an anecdote wherein curriculum needed to be updated. The reality is that unless there is an outside exigence, even the most insignificant curricular decisions can take years. While all programs must abide by the processes of their institutions, it's almost always possible to expedite or work around some of those processes. However, keeping curriculum current should be an ongoing activity—likely closely related to the program's assessment plan. Like most of the suggestions here, when the cuts come, it's most likely too late.

Make sure programs match student and local needs and are in line with your institutional mission. This should be a given. However, too often programs match faculty interests and tradition rather than changing student and external needs. If your program already offers a degree in technical communication, for example, make sure the degree requirements are current and address what students will need once they graduate. Requiring courses that give an overview of technical communication is a good thing. Requiring courses that speak to a specific technology that may be or is already outdated makes no sense—yet many programs do that. I'd recommend that courses be general enough that they can be regularly tweaked by changing readings and assignments without creating new courses. Of course, while new courses may take a long time to get approved, it still pays to offer them as special-topics courses before final approval. I can't overemphasize that market needs—both student and industry needs—need to be met more than faculty needs, especially at the undergraduate level.

In order to assure that curriculum meets current needs, I suggest implementing a real and ongoing process of curricular review. I would also strongly recommend that this curricular review be tied to program assessment. Information gathered from assessment can help programs determine the relevance of current curriculum. On the front end, most simply, all new courses should show how they meet program outcomes. In addition, a brief needs analysis would help assure that new courses meet more than individual faculty wants.

Programs should also be flexible in method of delivery. Today that usually means online as opposed to face-to-face. However, a variety of hybrid options are also possible. If your student population is local but tends to have full-time jobs, hybrid courses, with the few face-to-face meetings during nonworking hours can be successful. Likewise, *where* the courses are offered can make a difference. Sometimes, especially when getting to campuses means driving a long distance, especially right after work, offering classes at off-campus sites that are closer to workplaces is a good way to attract more students. A corollary of this, depending on the industries in your area, is to work with one large employer that may have a group of employees who might benefit from an applied writing class to offer a special section on-site for their employees. In most instances, a course like that tends to be a generic technical communication or business communication class—much like service classes. However, successfully offering such a class can be a great tool for recruiting new students into the program as well as helping to create an industry advocate.

Since students in applied writing programs tend to be nontraditional, programs need to be flexible in understanding that school is just one part—not the only part—of a student's busy life. That doesn't mean cutting students slack in what the program and the courses require the students to do but rather for the program to be flexible in offering courses when and where students can take them and being reasonable in things like accepting transfer hours.

One of the truisms about any successful program is that there's never enough faculty. This may become even more true during our new "austere times." Full-time, tenure-track faculty are seen as a luxury. Yet, for the time being at least, it's impossible to offer classes without faculty. There's a fine line between always having a demand, which is good, and consistently not offering enough classes, which can be construed as mismanaged. When administrators are looking for programs to cut, programs viewed as mismanaged rise to the top. The thinking is simply if a program doesn't care enough to run itself well, why should anyone else?

In order to avoid becoming a target because your program isn't managed well enough to offer sufficient sections of classes, it's important to have a method to ensure that there are enough faculty to staff those sections. The obvious answer to the staffing problem, and a solution writing programs of all kinds have used for years, is to hire part-time faculty. There are, of course, all kinds of potential problems with part-time faculty. However, part-timers, when chosen well and trained (an idea that too few programs consider), can become a program asset. The model for finding part-time faculty that most writing faculty know is what is commonly used for First Year Composition (FYC). That usually means that anyone with a master's degree in English or English education can teach the course—even if those people have no training or expertise in writing. That's a bad model and shouldn't be followed by applied writing programs. Programs that have master's programs in applied writing and have graduates working locally as practitioners probably have the best pool from which to find part-time faculty. Most programs don't have that luxury. Still, most programs should be able to find practitioners with some kind of master's degree. Honestly, I think that it is more important if the individual is a practicing writer and is thoughtful about writing in more than the area of their master's degree. People who write for a living often know more about teaching writing than they realize. However, just throwing them into a class with no training is often asking for disaster. It's crucial that some kind of training and mentoring is needed to help new teachers acclimate. The good news is that unlike FYC, wherein there are often many new part-timers every year, there are usually only a handful of new applied writing part-timers.

Connections

Perhaps the most important thing that can prevent programs from being targeted for cuts or extinction is having the right connections. Having connections doesn't mean just knowing the right people both within and outside of your institution but rather being important to others. It means that if what you do is reduced or goes away, others will also be significantly impacted. If cutting your program means hurting others who are important, your program then is more likely to be too important to cut.

Internal Connections

The easiest place to start with internal connections is with a program's service courses. In fact many applied writing programs seem to exist primarily to deliver service courses—usually to help engineering or business programs meet their accreditation needs. Traditionally, academics have tended to treat service courses with disdain. Frankly, that's a bad idea, especially for applied writing folks. Service courses can easily be seen as a reason for being. However, in order for service courses to be something more than simply a required course your unit delivers for another unit, you have to actively cultivate relationships with the units your courses serve. Too often applied writing service courses become generic technical communication or business communication courses offered in isolation from disciplinary curriculum. When there is little or no connection between the service writing course and the rest of the disciplinary curriculum, "austere" times will often see institutions exploring different solutions to meet accreditation needs.

The best way to avoid losing a service course is to make the course and its curriculum vital to the unit being served. Applied writing faculty need to reach out to disciplinary faculty, much in the same way a good Writing across the Curriculum (WAC) or Writing in the Disciplines (WID) program works. Working with the disciplinary faculty on the kinds of writing they want their students to do, both while students and then as professionals, goes a long way in making sure the service course curriculum is relevant to both faculty and students. While informal meetings might be a good way to begin, programs should think about some kind of regular curricular meetings between writing and disciplinary faculty. Formal curriculum committees with both writing and disciplinary faculty may be too much to expect; however, a once-a-term meeting to evaluate curriculum and suggest curricular changes is almost always a possibility. Keeping curriculum fresh, making sure as it is assessed that it meets the outcomes of both programs, goes a long way in keeping the target off our back.

While programs in engineering and business have historically provided the biggest need for applied writing service courses, don't think those are the only two programs that might benefit from an applied writing service course. The more units you deliver service courses to, the more valuable you become to the institution as a whole. Service courses for engineering and business tend to be driven by accreditation requirements. However, those two disciplines are not the only ones whose faculty understand the need for their students to have good written communication skills. While academic writing needs are often addressed by good campus WAC or WID programs, faculty in professional disciplines are also more likely to see the need for their students to have instruction in the kind of written discourse that they'll need in the workplace. Public administration and criminal justice immediately come to mind as possibilities. In addition, most campuses that offer degrees in the health professions have a real need for writing courses designed for those particular programs.

For example, several years ago, the nursing program at ASU approached the technical communication program about developing a service course for one of their majors. Consulting with the nursing faculty, the technical communication faculty developed the course. We now meet at least once a semester to see how things are going and what might be needed. Originally, the course was designed to give students some experience working with the kinds of writing assignments they might need to write in the workplace. As we've progressed, the nursing faculty still like that idea but also want to stress academic writing for nurses. It's an interesting tension because each course can contain only so many assignments. What we're seeing emerge is that more of the nursing faculty are beginning to understand that one writing class can do only so much. They're also beginning to realize that if they're going to assign writing in their own courses, they need to understand more about how to construct assignments in order to have their students produce the kind of results they're looking for. Ideally, in less "austere" times, we expect there would be a strong possibility for establishing some kind of WID program within nursing. Right now, that's impossible. However, the connection between nursing and technical communication keeps getting stronger. They realize that without our course, and the way it connects to their program, their program's value would be reduced. It's also important to note that the connection between nursing and technical communication is not just happening at ASU. Other institutions (Eastern Michigan quickly comes to mind) are teaching similar courses.

External Connections

Historically, institutions of higher education, especially individual academic programs, didn't think much about what we now call external stakeholders.

However, since higher education now sees all of those external stakeholders as potential revenue streams, everyone is interested in them. There's absolutely nothing wrong with an external stakeholder giving your program money (unless, of course, there are problematic strings attached). More important for the survival of your program than even money is that those stakeholders become strong advocates for your program both to your administration and in the wider community.

Interns

One of the more traditional and easiest ways for applied writing programs to gain external ties with the community is internships. Programs with established internships should do what they can to maintain them and enhance them. If the internship coordinator does not make regular visits to the site, that should change. Site visits not only give faculty coordinators a better understanding of the environment their students are experiencing, it also is a wonderful opportunity to strengthen connections between the program and both the individual site supervisor as well as the sponsoring organization. Being more visible and strengthening the relationship can not only lead to more internships but also to building external allies.

There's the proverbial good news and bad news if your program does not have regular and established internships. The good news is that with the new "austere" philosophy, businesses are looking to do things "on the cheap." Interns are considered cheap. The bad news is that it's still difficult to find organizations that will take interns. Since more and more intern projects, like much work, are project-based, when an organization does look for an intern, it is only to do one project. That means that new internships are often one-time situations. Maintaining internships can be even more problematic when students have a desire to remain in the community where they go to school. (Since many applied writing students are nontraditional, they likely already live in that community.) If an internship works well, the student finishes the internship with a full-time job offer. I've seen this happen time and time again, even during the most "austere" of times. That's great for students who become gainfully employed. However, it means the program needs to look for more internships.

Yet a student who finds full-time employment as the direct result of a successful internship becomes doubly valuable to the program. Over time, and surviving is always about long-term thinking, those former students then become more likely to encourage their employers to hire more interns. At both institutions where I've supervised interns, I've seen past interns advocate for new interns, creating a multigenerational intern culture.

Alumni

Keeping connected with alumni should just be a normal part of how institutions and programs within those institutions do business. While there is a strong alumni culture at some schools, usually schools with well-established traditions, many schools both at the institutional as well as program level do a terrible job of keeping connected with and using alumni. No matter what the institution may do concerning alumni relations, it's imperative for applied writing programs to keep track of and keep in contact with their alumni. As I've already mentioned, alumni can lead to internship possibilities for the program. They can also provide real job leads to new graduates. Keeping in touch with local alumni can provide opportunities for the alumni to give presentations for students about some of the projects and workplace situations the students might face in the future. Those presentations might be classroom-based or they can be held in the evening for the whole program—students and faculty.

In the past, just keeping track of alumni could be a huge problem. However, now email addresses might suffice. Emailing alumni every so often about what's going on in the program keeps them in touch. Obviously, having a program Facebook page seems to be a necessity. However, I'd also strongly recommend spending just a little money and having a program Ning (or some other private social network) site. Having an active social network for the program connects alumni, current students, and faculty. I'd also recommend highlighting the strong alumni relations in all program annual reports and other such documents that filter upwards. Again, when administrators are looking for targets, they first look for the easy ones. Programs that have an acknowledged strong and active alumni base don't make easy targets.

External Boards

External boards are a good way to connect multiple external communities to the program. There's no one good single factor in determining who should serve on a board. However, the broader based a board's membership is, the more influence it might have within the institution. It's easy to think about having a board composed of practicing writers, perhaps an alum or two, and the like. That's a good start. However, since membership on almost any board tends to confer status on the members, I'd suggest programs aim high. Think about inviting someone from the executive level, in some instances even a CEO, of a local business that employs writers. Think about the Executive Director of a large local nonprofit. Since the board is composed of members external to the program, one or two university personnel might also be possible. If your program has a service course for engineering or business, I'd seriously think about inviting the Dean of Engineering or Business to serve on the board. Once the program faculty starts thinking along these lines, the local possibilities begin to emerge.

While it's nice to have a board, it's better to have a board that's active. The people who might serve on your board will be busy, but they'll expect that they've been invited to serve because they have expertise. You need to use their expertise. Again, your local conditions might dictate some obvious uses.

However, if you're just beginning with an external board, you might start with presenting your curriculum and outcomes to them with the rationale for both. (If you can't easily come up with a rationale for outcomes or curriculum, that should be an indicator that creating understandable and sensible program outcomes and a curriculum that is sensible and not just thrown together is the first thing your faculty needs to do.) Once the board is familiar with the program and its outcomes, then it makes sense for a discussion on how the program meets external needs or what might be done to have the program better meet needs. It also makes sense to have members of the external board be involved in some kind of culminating project evaluation. Many programs require a senior capstone project or program portfolio, having members of the board serve on a committee reviewing those projects not only gives the board members a sense of participation and gives their perspective on your students' work, it also creates a real external audience for your students. The reality is that any time you can make your students' good work public, you are strengthening your position and making the target less likely.

As we've seen, there are ways applied writing programs can actively work to be relevant to their students as well as internal and external constituencies. To do so they must be flexible and make sure they have strong connections both within and outside of the institution.

POTENTIAL PITFALLS

While I've tried to be positive and have written about creating an environment that will allow programs to be sustainable, the reality is that there are always some situations that clearly make some programs targets. The single biggest problem that I think exists for applied writing programs is that they are applied programs and not part of the traditional academy. What makes this an even more dangerous situation is that, more often than not, they are buried within traditional English or humanities departments whose cultural and disciplinary values are diametrically opposed to the worldview of applied writing faculty. Even when there is a large rhetoric and composition contingent within the unit, the rhet/comp faculty often tend to align themselves with their humanities-based colleagues. The reality is that when you don't fit, you become a target. While having an independent program doesn't guarantee anything (as my program at ASU can attest), it does make things more difficult. It's no secret that I think independent writing programs should be independent departments. While I've noted that some rhet/comp faculty in English departments may lean toward the humanities, they're much more likely to support their applied writing colleagues when there are only writing faculty in the unit.

I've previously mentioned that it can be a good idea to use some part-time faculty. In "austere" times, however, it is dangerous to have administrators think entire programs can be delivered with one or two full-time faculty and dozens of part-time faculty. I think the best way to limit the number of part-time faculty is the "expertise argument." While your institution may still believe that anyone can teach FYC, you should create an environment in which everyone knows that not everyone can teach an applied writing course. The criteria you use to hire your part-time faculty, how you train them, and the professional development activities you require from them defines this argument.

Again, being flexible in how you deliver your classes and using online delivery methods is a good thing. However, don't blindly assume that all online is good. The best practices for online delivery are similar to the best practices for face-to-face classes. Getting students involved and interacting is good. Using the technologies to engage students in making the most out of the online delivery is also good. However, just using course management systems as a place to store tired lectures and having students respond to age-old assignments doesn't work any better online than it does face-to-face. Likewise, the current fascination with MOOCs (massive open online courses) is something to be wary of. That's not to say that using digital technologies to deliver some information to large numbers of interested people is a bad idea. What is crucial is whether it's effective in delivering an applied writing curriculum. If you think that it's unlikely that anyone could think of delivering a technical communication service class using a MOOC, I'd suggest you think again. Administrators have a long history of making what they think are good decisions by being deluded with bad uses of new technologies. If, for example, Georgia Tech or MIT decided to offer a technical communication service course via a MOOC, might your Provost and Dean of Engineering think that might be a viable option?

MOOCs might never become a threat to any applied writing program. What's important in thinking about MOOCs is that doing so gives us a way to think about maintaining a sustainable future. Identifying potential problems—analyzing what they do or don't do and how some of those functions might enhance what you want to do—means your program is prepared to respond to potential cataclysmic changes.

FINALLY . . .

Perhaps the best way to end talking about building a sustainable applied writing program may be to tell a story that can help us navigate any fast-changing environment. A long time ago, when I was first learning to drive, I can remember that during one of my first experiences driving on a freeway, my driving instructor told me that I needed to drive where I'm going—not where I am. If I see a curve, to turn when I see it—not to wait till I get there. What he was trying to explain was basic physics. When you're traveling at 65 mph, you need to anticipate where you're going—not wait until you're there to act. If you don't, you'll run off the road. His advice holds for more than driving. While we too often remember where we've been and focus on where we are, if we're to continue, we must always understand where we're going. We must also understand that we may not be the ones who build the roads, but we are responsible for whether we get to where we want to go.

REFERENCES

Greene, J. P., Kisida, B., & Mills, J. (2010, August 26). Colleges feeding administrative bloat. *Atlanta Journal-Constitution*. Retrieved from http://www.ajc.com/news/news/opinion/colleges-feeding-administrative-bloat/nQjgt/

Maid, B., & D'Angelo, B. J. (2013). What do you do when the ground beneath your feet shifts? In K. Cargile Cook & K. Grant-Davie (Eds.), *Online education 2.0: Evolving, adapting, and reinventing online technical communication* (pp. 11–24). Amityville NY: Baywood.

Samuels, B. (2009). Corporatizing the instructorate: Ceding authority to the administrative class. *College Composition and Communication, 61*(1), A8–A10.

White, E. (1999, January 11). *Increasing state intervention.* Message posted to WPA-L.

CHAPTER 2

A Response to Austerity: Using Ecopreneurship to Build a Sustainable Writing Major

Amanda Bemer and Teresa Henning

This chapter makes the somewhat optimistic argument that even in times of austerity, writing program development can be rewarding if one is willing to be an "ecopreneur" (Ivanko, 2008) who is committed to make a living while supporting the health of others through selective use of economic and sustainable practices. Ecopreneurs find a way to support themselves by being "creative," "flexible," "freedom minded," and "risk tolerant" enough to "not only transform the landscape but coalesce into a movement to transform global problems into opportunities for restoration and healing" (Ivanko, 2008). Using our own experiences developing an undergraduate professional writing major at Southwest Minnesota State University (SMSU), this chapter will explore the ways ecopreneurship played an important role in starting a new writing major and sustaining its growth over the last 4 years. Part of this exploration will include some discussion as to why it made sense for our institution to enlarge the major beyond technical communication to include the broader category of professional writing.

To understand the exigency of this argument, it is important to consider the austere constraints under which the undergraduate Professional Writing and Communication (PWC) major operates. SMSU is a small, rural, liberal arts, 4-year residential and commuter campus that is part of the state funded Minnesota State Colleges and Universities system. SMSU is the sixth-smallest of the seven universities in the system, and in this time of economic instability has felt the consequences of budget cuts more keenly than other, larger campuses. Given these facts, we feel lucky that the PWC major is growing even while we are concerned by our lack of resources.

Starting a program in 2008 during a time of austerity has had several troubling consequences for us. First, while we have two faculty members (the two authors of this chapter) who teach in the program, neither of us has reassigned time to direct the program. Second, austerity has made our campus so FTE driven that classes with fewer than 12 students are cancelled as are programs with consistently low-enrolled courses. To avoid starting a program that could not get off the ground due to cancelled courses, nearly the entire PWC major has been developed from existing courses that serve other majors.

Further, we have no space for our major—literally. Unlike established majors who have program office spaces, display cases for student work, and student lounges, we have only our two, cramped, dated, faculty offices from which to work. Moreover, our access to adequate technology is limited. Our students do not always have access to the software they need let alone computer classroom space. On a campus with only five computer labs, each with approximately 24 seats, obtaining a computer classroom is a premium request that is not always granted. In order to get computer space, we have to be creative with class scheduling.

Essentially, all of our technology and technology space is controlled by others who have no personal interest in the success of our program—because of this, we are constantly begging for resource sharing and are somewhat invisible to others on campus—we have no visible space (or mark), so it is very easy for students, faculty, and administration to forget about us.

Given these constraints, it is easy for us to become frustrated, and we often worry about our ethos—how are we to run a program when we constantly have to beg for resources? How are we to be taken seriously by students, parents, and employers if our own institution cannot support the work we do? The figure of the ecopreneur is especially attractive as it suggests a way for us to take charge of some of our own ethos. To further develop this point, this chapter will consider how strategies related to ecopreneurship might be pertinent to conversations about technical communication, austerity, and growth. Specifically, this chapter will consider how our experiences relate to systems thinking and sustainable writing program administration, the importance of local context in ecopreneurial strategies, and how we use these strategies for growth. Finally, we conclude with the risks and rewards ecopreneurial strategies can have for program development.

SOME FEATURES OF SUSTAINABLE PROGRAM DEVELOPMENT

Before considering how ecopreneurial strategies relate to sustainable program development, it is important to reflect on the key features that make a writing program sustainable, such as using a systems-thinking approach to design a writing program that is an open system and strategically crafting a technical writing program definition by paying careful attention to relationships.

Using Systems Thinking to Design an Open System Writing Program

We assert that sustainable writing program growth requires a systems-thinking approach to program administration. We agree with Ashe and Reilly (2010) that "imagining our universities, departments, programs, students and faculty as part of an academic ecosystem has both utility and precedent" (p. 92). Like Ashe and Reilly, we find that taking a systems-thinking approach allows us to attend to relationships and resources.

Like Ashe and Reilly (2010), we are also aware that viewing a writing program as a system has a historical precedent in our field starting with Cooper's "The Ecology of Writing," which Dobrin and Wiesser (2002) develop further in their book *Natural Discourse: Toward Ecocomposition.* We are especially attracted to Dobrin and Weisser's definition of ecosystems as "groups of organisms which function together in a particular environment (physical and chemical) and exchange energy within the system in order to metabolize, grow, and reproduce" (p. 73) and agree with Cooper (1986) that "systems that constitute writing and writers are not just like ecological systems but are precisely ecological systems" (p. xiv). The ecological perspectives these scholars provide are especially useful for discussions of space, geography, and relationships (Ashe & Reilly, 2010, p. 93). In addition, such perspectives allow one to bring to bear work in environmental science about ecosystems, which reminds us that healthy ecosystems are diverse spaces that value both competition and cooperation as means of making sustainable use of resources (Fiksel, 2003, p. 5331).

However, Fiksel (2003) points out, not all systems are healthy and capable of resisting decay. A system's health depends upon its resilience—its ability to "resist disorder" (p. 5332) by acquiring energy. Healthy systems are diverse, adaptable, efficient, and cohesive; "they continually draw upon external sources of energy and maintain a stable state of low entropy" (Fiksel, 2003, p. 5332).

For us, an open system resembles a kaleidoscope. In Figure 1, we have sketched our vision of an open system. Embodied within a large circle denoting environment is our open system (i.e., the PWC program). The circles that overlap the outside of the open system represent stakeholders—these resemble a kaleidoscope because they have the ability to move. Stakeholders for an open system, particularly ours, are somewhat transient. New stakeholders can be added; old stakeholders can leave. They have varying levels of interest in the system, and this interest can shift over time. The open system is fed by energy represented as black dots in Figure 1. These dots are throughout the image because energy (e.g., students, new professors, money) is a lot like light—it

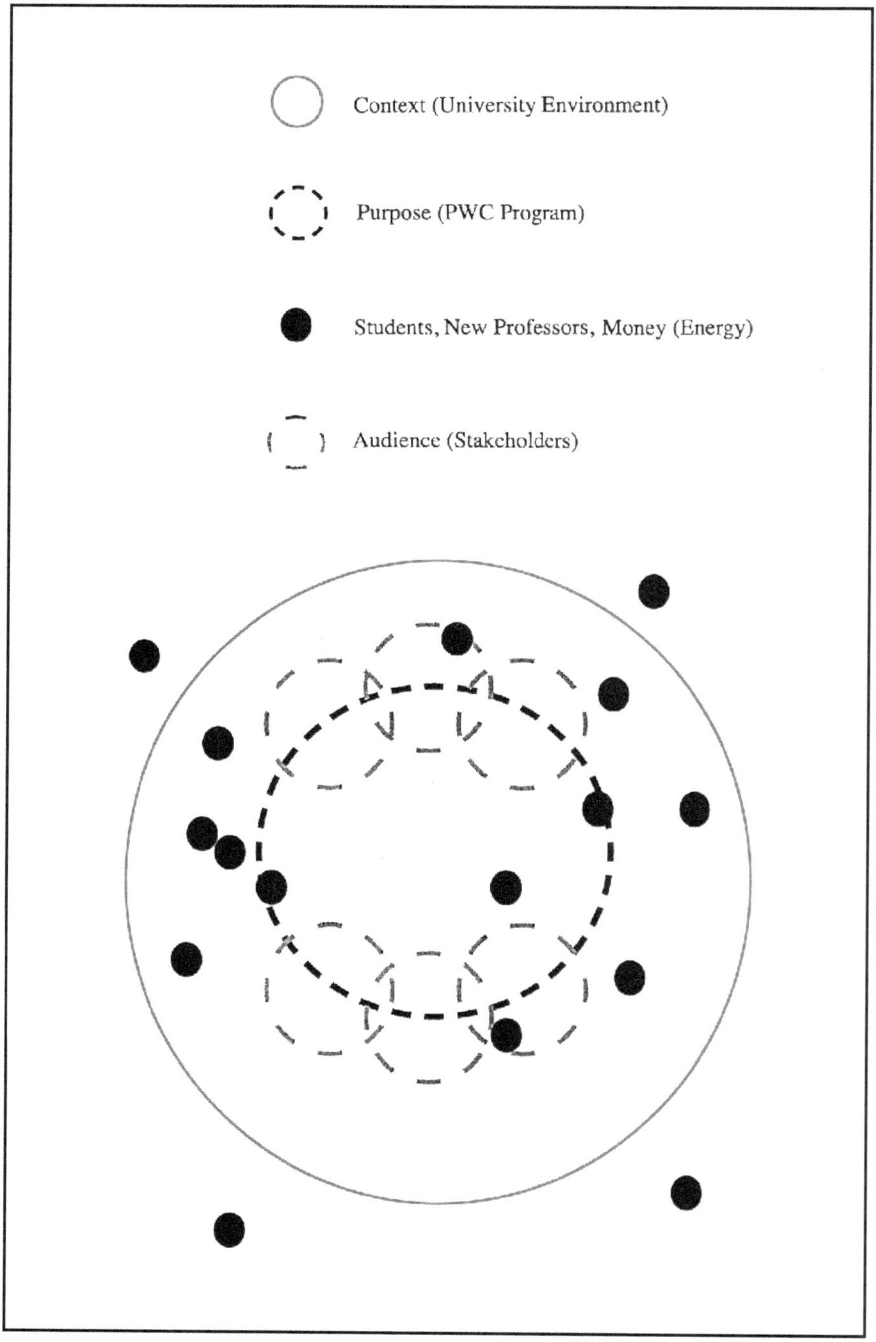

Figure 1. Our vision of an open system.

comes to us from many directions and places, and without this light (energy) there would be no (visible) system in the first place. The system takes this energy and transforms it into our output—graduates with strong writing skills.

Thinking of an open system as a kaleidoscope emphasizes the vital role that relationships with others play in a system's resilience. Fiksel (2003) argues that "a sustainable enterprise is one that continues to grow and adapt in order to meet the needs and expectations of its shareholders and stakeholders" (p. 5331). Being sensitive to the needs of shareholders and stakeholders is also a feature of developing a sustainable writing program. New programs change an ecosystem; changes in the environment of a department (classes offered, programs created or killed) affect every member of the department (Ashe & Reilly, 2010, p. 93). Thus, creating space for a program also involves building strong relationships that foster interdependence.

Warnick, Cooney, and Lackey's (2010) discussion of the difficulties of sustaining a writing studio program suggests that for a program to thrive, faculty must have a stake in it. A program that is "embedded" within something that already exists (such as a class) is more sustainable than a free-standing entity because faculty already have a stake in it, and therefore faculty can more easily see how they would benefit from its success (p. 94). Warnick et al. thus contend that building strong relationships is a pathway to building space for a program leading to increased buy-in for program courses (p. 94). Buy-in can also be achieved through taking advantage of existing course offerings (Ashe & Reilly, 2010, p. 96), something that is possible for an interdisciplinary writing program.

Crafting a Technical Writing Program Definition That Is Sustainable

Taking advantage of the interdisciplinary nature of a writing program, however, requires paying close attention to the definition one uses for both the field and the particular major; this issue is particularly relevant for a program in technical communication because scholars, over the years, have discussed the difficulties of assigning a definitive label to the work of technical communicators. Understanding the limiting and dividing aspects of definition provides an impetus for a careful stance toward defining a program in the university setting.

For instance, it has been argued that the field of technical communication suffers from an "identity problem" that should be solved with an authoritative definition (Jones, 1995, p. 568). Some go so far as to say we have a lack of "brand identity" that prevents our field from having a "compelling reason to choose this service over others" (Carliner, 2012, p. 59). This identity problem can be seen in the way that professors of technical communication have had to consistently define and justify their area of expertise to other academics (Jones, 1995, p. 568). The broad nature of research in technical communication (Blakeslee & Spilka, 2004, p. 91) may make the academic field somewhat hard to classify and can hurt the sustainability of a program through its seeming relevance (or lack thereof) to other academics. Others have also posited an authoritative definition of technical communication as troubling because the "constantly evolving" nature of technical communication (Ecker, 1995, p. 570) makes it hard to create a definition that would not serve to harmfully limit the field's growth (Allen, 1990, p. 75).

This issue of definition has clear effects on programs, especially during the program creation process; writing faculty who describe the creation of an undergraduate writing major explain that simply coming up with an acceptable definition and programmatic space for the major is fraught with institutional tensions that can sometimes be insurmountable. Wallis May Andersen (2010) notes that writing faculty at Oakland University had to "overcome second class citizenships" (p. 75) in their efforts to have colleagues accept the fact that writing courses can do more than remediate a perceived lack of writing skills in students. Lowe and Macauley (2010) discuss how a similar situation at their former institution all but destroyed a new writing major and caused both authors to seek employment elsewhere. Ashe and Reilly (2010) and Langstraat, Palmquist, and Keifer (2010), among others, explore analogous tensions that can ensue when a professional writing major is developed and housed in a literature-centric English department.

Baker and Henning (2010) express similar concerns over the ways definition can lead to impoverished notions of composition and technical writing when such writing is defined "as a set of skills divorced from context" (p. 154). They explain how SMSU's professional writing and communication program with its roots in rhetoric has allowed them to resist limiting definitions of writing by invoking "all elements of the communication triangle . . . engag[ing] with and interrogat[ing] different historical perspectives of rhetoric . . . [and] . . . focus[ing] on the situational nature of writing, rhetoric, and program development" (p. 154).

By relying on a rhetorical definition of professional writing, the SMSU professional writing program is an open system that seems fairly resilient. Its interdisciplinary nature allows for diversity and a sharing of space with other courses and programs, thereby fostering good relationships with stakeholders and shareholders and creating efficiencies leading to interesting innovations that keep the program resilient. The next section discusses how using ecopreneurial strategies have helped us develop some of the efficiencies and relationships important for a rhetorical professional writing program that functions as an open system.

THE IMPORTANCE OF LOCAL CONTEXT WHEN APPLYING ECOPRENEURIAL STRATEGIES

To understand how ecopreneurship has allowed us to continue developing our program in a resilient manner, it is first important to define ecopreneur and consider how that term relates to an open system that values relations with others. When tied to systems thinking, this term can be an especially powerful concept for writing program administration. On one hand, the first half of this term allows for a focus on how an ecosystem's health is tied to its diversity and ability to foster cooperative relationships. On the other hand, the idea of earning a living in a creative manner, which is captured by the second half of the term, allows one to consider the ways competition can be a positive aspect of sustainable growth. Ecopreneurs innovate, often relying on what Newcomb (2012) describes as "situational creativity," which is the finding of "innovative ways to respond to specific situations" (p. 608), thus coupling context with innovation. The context-sensitive ecopreneur has the potential to synthesize spaces, relations, and disciplines in a manner that allows for creativity, diversity, cooperation, and competition. One can understand this potential by considering how the strategies of ecopreneurship relate to our institution's experiences in negotiating a space for the PWC major.

Our Local Context

In order to consider how ecopreneurial strategies have been useful in the development of what we hope is a sustainable writing major, it is necessary to offer some background information about SMSU's PWC major. As noted earlier, SMSU is a small, rural, liberal arts 4-year residential and commuter campus of about 3,000 students and is best known for its programs in culinology, business, environmental science, and creative writing. When Henning was hired for a tenure-track position in professional writing in 2006, the English department was home to three majors: a literature/creative writing major with a program enrollment of 35 to 50 students each year; a literature major with a program enrollment of 6 to 12 students a year; and a communication arts, secondary education licensure with an English emphasis major with a program enrollment of 6 to 12 students a year. It is also worth noting that while program enrollments of 6 to 12 students may sound small, many courses in both the literature major and the secondary education licensure major are courses required by the literature/creative writing major. In addition to offering these courses in 2006, the English department was also responsible for service courses in first-year composition, advanced composition, and scientific and technical writing.

By the end of her first year in the department, Henning gained approval for the PWC major at the departmental, university-wide, and system-wide levels. The major was officially listed in the catalog as of January 2008. The PWC major at SMSU is composed of six key areas, including written communication, oral communication, visual communication, history and theory, professional contexts, and professional expertise. As such, the major includes within it courses in graphic arts, speech, business, and ethics, not to mention courses in advanced composition, journalism, copyediting, new media, technical writing, and business writing. The major graduated eight students in May 2010, all of whom have jobs or funded graduate assistantships at other institutions. The major is now the second largest in the English department.

Relationships That Existed Before Applying Ecopreneurial Strategies

Making such a sudden change to the English department ecosystem was a process fraught with tension. While the department approved the 2006 new-hire in professional writing, it did not share the same vision as to what that person's duties should be. Some of the literature faculty were of the opinion that the line's primary concern should be to teach composition, by which they meant first-year writing. The creative writing faculty, however, were eager for Henning to develop the PWC major so that they could move the journalism courses out of creative writing and into the new major. Those on the hiring committee and the other compositionists were eager to see the major develop for both the sake of the students interested in the program and for the opportunity it would create to add variety to their own positions.

Sudden changes to a department's ecosystem are not only fraught with tension but they can also be risky. A lone organism cannot create or survive such a change. Survival depends upon sharing with others. In this instance, Henning found it prudent to build upon existing alliances by forming a Professional Writing Advisory committee on which a cross-section of tenured and nontenured colleagues were invited to serve. The committee became a sounding board for curriculum designs and rationales and for improving relations with literature faculty in the department. At the committee's suggestion, Henning consulted with key faculty in literature, business, science, and art to develop courses to serve their majors while also serving the PWC major.

Developing Relationships Affecting Ecopreneurial Strategies

These initial collaborations were crucial to getting the major approved and were useful in bringing the major to the attention of other faculty on campus. As a result, in 2010, the program was able to make an argument for hiring another professor to share in the teaching of PWC courses. This new line was somewhat of a compromise, however, as the administration was not willing to fund a tenure-track line for such a new program. They did agree, however, to what our institution refers to as a fixed-term position—a full-time position with benefits that is renewable as needed. This position has remained fixed-term for 3 years. Each year, the English department chair negotiated with the administration in an attempt to have this position converted to a tenure-track line. The tenuous nature of this position has made us quite sensitive to the health of our major. And, we are happy to say, that in the time it took us to write this chapter, Bemer's position will become a tenure-track one in the fall of 2013.

HOW WE USED ECOPRENEURIAL STRATEGIES FOR GROWTH

By establishing space for the major and creating thoughtful relationships with other stakeholders, we have been able to create a major that functions as an open system. As an open system, the major lends itself to some interesting and ongoing ecopreneurial bartering, rescuing, reusing, exchanging, recycling, sharing, and repurposing, strategies that are all helping us to develop the major in a sustainable manner. To illustrate this point more fully, we will consider three important barters that grew out of collaborations with art, science, and SMSU's liberal education program that have led to some interesting and ongoing bartering, rescuing, reusing, exchanging, recycling, sharing, and repurposing, which have helped the program develop in a flexible and sustainable manner.

Bartering Can Support Collaborating, Rescuing, and Reusing

While working with the graphic arts professor to develop the visual core courses in the PWC major, Henning gained access to a dedicated lab with the entire Adobe Creative Suite in exchange for helping him and a marketing professor write a new interdisciplinary minor that would pull together graphic design courses, marketing courses, and writing courses to create an advertising emphasis.

In addition to successfully bartering with the graphic arts program, work on the new minor allowed for both rescuing and reusing by tying the PWC major's upper-level, low-enrolled Writing and New Media course to the new minor. This course, with its emphasis on social media, blogs, and Web pages, is a good choice for an advertising minor, to which graphic arts and marketing colleagues readily agreed.

Not only did this barter invite rescuing, but it was also a good opportunity to reuse some of the assessment work completed when developing the PWC major. For the minor's writing courses, Henning was able to reuse the learning outcomes and planned assessment strategies, thus modeling for the professors how to write their sections of the minor proposal. In the end, bartering with the graphic arts professor has meant that faculty members from three disciplines have built alliances with each other. As a result, students across disciplines have benefited by our willingness to share space by gaining access to technology and an interdisciplinary minor in advertising.

Collaborating Can Lead to Bartering, Sharing, and Repurposing

As was true of the initial collaborations with the graphic arts professor, collaborations with science faculty yielded three resources from which PWC majors benefit: the campus's Natural History Museum, which the science department runs on a shoestring budget with no compensation for the director who volunteers his or her time; a GIS lab with a tablet printer capable of printing full-color posters for research conferences; and an Undergraduate Research Conference that science also organizes. Henning worked out a barter with science that for one of its projects, the scientific and technical writing class would create a set of exhibits for the museum, if the science department would allow some limited printing of posters on their GIS printer. The original museum exhibits took the form of technical description posters describing such concepts as wetlands, fish species of Minnesota, and prairie plants of Minnesota. Since then, the museum received a small grant for flat screen televisions, so students in scientific and technical writing now create scientific, multimedia videos for the museum as one of their projects.

In addition to these resources, science has agreed to share conference space with other disciplines if interested departments share in the cost and administrative work of setting up the conference. Each fall, the scientific and technical writing class now presents at the conference, and Henning or Bemer, along with 10 other faculty, serve on the Undergraduate Research Conference committee and donate a small piece of our department budgets to the conference.

Tying scientific and technical writing to both the Natural History Museum and the Undergraduate Research Conference has allowed the course to provide the research skills PWC majors need while also bringing some notoriety to the work of students through conference research awards and articles in the local newspaper, thus rescuing the teaching of writing from second-class status.

Context Sensitivity Allows for Sharing and Building New Relationships

As well as using ecopreneurial strategies to develop relationships with colleagues in art and science, we have had opportunities to develop relationships with colleagues across campus vis-à-vis the campus's new liberal education program. At the time of the development of the PWC major, the entire campus was beginning the process of revising its liberal education courses. As part of this process, the English department's composition committee, on which both Bemer and Henning serve, began redesigning its first-year composition program and proposed requiring four writing courses that students would take over the course of a 4-year degree.

Although it is beyond the scope of this chapter to detail all the challenges of writing program development done only by committee and without the benefit of a WPA, it is worth noting that after much compromise, SMSU students are now required to complete four writing courses. However, only the first-year course is required to be an English course. Given the limited resources of most departments on campus, in practice the English department has been providing the first- and second-year writing courses while faculty in other disciplines across campus provide junior and senior writing courses in their majors.

Faculty choosing to use a new or existing course to meet the junior and senior writing requirements for their major must receive approval from the liberal education curriculum committee as well as the full faculty assembly. The PWC major uses its scientific and technical writing course as its junior-level writing course. It, like all junior-level courses, went to both the liberal education and faculty assembly committees for approval. This process brought the course to the attention of both the computer science and exercise science faculty. The end result is that both now require this course as their junior writing course, and we now offer two (rather than one) sections of this course each semester.

THE RISKS OF USING ECOPRENEURIAL STRATEGIES

We close by considering the risks and rewards associated with using ecopreneurial strategies to develop a writing program so that it functions like an open system. Shamoon, Schwegler, Trimbur, and Bizzell (1995) argue that when designing a new writing program, it is important to consider what is at stake (p. 14). We agree and would contend that this question should be asked again and again with each new development one plans for a writing program. Further, we agree with Baker and Henning's argument that writing program development always puts at risk our disciplinary integrity. We should then be asking not only what is at stake but also how those stakes affect our integrity.

Questioning how program development relates to one's disciplinary integrity is important because, as Baker and Henning (2010) note, these terms, when brought together, are potentially "agonistic" (p. 168). Disciplines tend to "discipline rather than encourage progressive practices" (Bullock, 2000, p. 24), while integrity "suggests a desire for wholeness or completeness" (Baker & Henning, 2010, p. 169). Keeping these terms in balance so that we respect the needs of our stakeholders as well as our major as a whole is both what we risk and what we can gain when bringing ecopreneurial strategies to writing program development. This point will become more obvious as we consider the specific risks and rewards we have observed.

Building Relationships and Sharing Spaces Invites Risk

Integrity, ecopreneurial strategies, and open systems all place value on relationships with others. However, relating with others is a risky endeavor because not all relationships are beneficial to an ecosystem. From a purely biological perspective, invasive species pose a significant threat to an ecosystem because, rather than cooperating with others, such species dominate the ecosystem by killing off all other life forms. The end result of this kind of behavior is that the ecosystem becomes uninhabitable even to the dominant species because an ecosystem needs species diversity to stay healthy. Similarly, disciplinary impulses for homogeneity need to be tempered with concerns for diversity so that those disciplinary impulses do not harm other species and thereby destroy the health of the ecosystem.

However, it is not only our disciplinary impulses that must be tempered when developing an open writing program with ecopreneurial strategies. Our desires for wholeness and connection with others must also be tempered so that the relationships we develop with others are not one-sided. Recognizing that relationships are dynamic and can change is also important because even relationships that start out in a healthy manner can, without careful nurturing, deteriorate. Further, each member of a relationship is an actor in many different contexts that are also always changing and shifting, adding another layer of complexity, which means that even carefully nurtured relationships can still become one-sided. While it may seem contrary to the idea of making connections, at the start of a relationship, writing program administrators should think carefully not only about how to build a particular relationship but also think about what they will do should that relationship become unhealthy.

To better understand just how precarious relationships can be, it is useful to contextualize this discussion by considering an area of continual risk in SMSU's PWC major—the fact that nearly all of our upper-level writing courses are now also service courses. Essentially, we have attached our livelihoods to professors outside our department. We now must consult with other departments about when to offer our courses and how many sections to offer. We have to be sensitive to the fact that some might feel threatened by our courses taking FTEs that belong to other majors.

Further, we are also putting ourselves in the position of having to adjust our curriculum to meet the needs of other departments, adjustments that may not be in the best interests of our majors. A good case in point is our scientific and technical writing course. So far, this chapter has pointed out that sharing this course with science has given PWC students access to both the Undergraduate Research Conference and the Natural History Museum. Yet this course is also fraught with difficulties. First, the linkage between science and technical writing is difficult to make when only about a third in the class are science majors. Second, we worry that our majors seeking jobs in technical communications are simply not getting enough experience with technical writing from this course.

As a result, we have made some compromises. We have convinced science to let us drop the course's résumé and cover letter requirement to fit in usability testing and empirical research. For those PWC majors interested in technical writing, we work hard to get them involved in internships and taking courses online from a nearby, larger, 4-year institution with more offerings in technical writing. The risks with this latter compromise are that we have less control over what our majors learn about technical writing and end up supporting an institution that is in some ways a competitor. While we enjoy collegial relationships with the faculty members at that institution, we are fully aware that faculty move on, so we feel some unease tying our major to this other institution.

Staying Responsive to Change and Managing Resources Are Problematic

An ecopreneurial approach advises us to be flexible and responsive to change. In order to survive, ecosystems adapt to new species and changes to the environment. Ideally, a writing program should do the same—and we have been adapting the PWC program to fit changing needs. However, university environments are notoriously slow. Our institution has a unionized faculty, and every proposed new course must be approved by both a curriculum committee and then the faculty overall. This process is somewhat lengthy and requires foresight, so we must plan for curriculum changes in advance.

This slow adaptation can hurt the PWC program—failure to offer appropriate coursework might cost us prospective students. Of course, this process does allow us time to thoroughly consider all aspects of change to our offerings and how they may affect our relationships across the university, our environment. In the end, there does not exist a solution to this slow-moving nature of the university environment.

Ecopreneurial strategies also cannot fully address problems with limited resources. At the beginning of the chapter, we discussed how our austere circumstances limit our habitat. What to do when our habitat is out of resources and no resource can be repurposed to fill our need is a risk of ecopreneurship that we deal with often. We simply cannot barter for every resource we might need because some resources are just unavailable.

THE REWARDS OF USING ECOPRENEURIAL STRATEGIES

Even though it is clear that designing an open writing program with ecopreneurial strategies in a time of austerity is fraught with risks, we contend that these risks are more a function of writing program development itself and less a function of our chosen approach. In fact, our chosen approach, with its focus on contexts, relationships, and resources (i.e., energy), forces us to confront some of that risk with each programmatic decision we make. In some ways, this approach has made us less naïve. We have now come to expect that our relationships will change and new stakeholders and resources will have to be found. We also realize that not everyone will necessarily care about the health of our ecosystem in the same manner that we do. As a result, we are able to be more critically minded about what relationships we develop and which ones we avoid.

Overall, ecopreneurial strategies give us an optimistic framework for critical decision making, reminding us that stakeholders are transient so we do not overrely on particular relationships to sustain the program and helping us stay open to forming new relationships. Further, these strategies give us access to resources that we need for no or little cost (e.g., computer labs and publication forums.)

Additionally, the term "ecopreneur" itself fits well with our concerns about our disciplinary integrity. Like the terms "discipline" and "integrity," the word "ecopreneur" represents a potentially agonistic relationship between the ideas of protecting the health of an ecosystem by fostering cooperative relationships while making a living—a task that involves competition. Yoking together values of cooperation and competition yields perhaps the most rewarding benefit of this approach and that is that ecopreneurial strategies invite innovation via "situational creativity" (Newcomb, 2012, p. 608), making program development more interesting and enjoyable. Part of that enjoyment comes from the fact that we are more willing to take calculated risks, such as sharing space with other programs and other stakeholders. Moreover, we also enjoy the interdisciplinary thinking this approach invites. Finally, taken together, these benefits allow us to create a dynamic program that is responsive to change even in times of austerity.

REFERENCES

Allen, J. (1990). The case against defining technical writing. *Journal of Business and Technical Communication, 4*(2), 68–77.

Andersen, W. M. (2010). Outside the English department: Oakland University's writing program and the writing and rhetoric major. In G. Giberson & T. Moriarty (Eds.), *What we are becoming: Developments in undergraduate writing majors* (pp. 67–80). Logan, UT: Utah State University Press.

Ashe, D., & Reilly, C. A. (2010). Smart growth of professional writing programs: Controlling sprawl in departmental landscapes. In D. Franke, A. Reid, & A. DiRenzo (Eds.), *Design discourse: Composing and revising programs in professional and technical writing* (pp. 89–111). Fort Collins, CO: WAC Clearinghouse/Parlor.

Baker, L., & Henning, T. (2010). Writing program development and disciplinary integrity: What's rhetoric got to do with it? In G. Giberson & T. Moriarty (Eds.), *What we are becoming: Developments in undergraduate writing majors* (pp. 153–173). Logan, UT: Utah State University Press.

Blakeslee, A. M., & Spilka, R. (2004). The state of research in technical communication. *Technical Communication Quarterly, 13*, 73–92.

Bullock, R. (2000). Feathering our nest? A critical view within our discipline. In L. Shamoon et al. (Eds.), *Coming of age: The advanced writing curriculum* (pp. 19–24). Portsmouth, NH: Heinemann.

Carliner, S. (2012). The three approaches to professionalization in technical communication. *Technical Communication, 59*(1), 49–65.

Cooper, M. M. (1986). The ecology of writing. *College English, 48*(1), 364–375.

Dobrin, S. I., & Weisser, C. (2002). *Natural discourse: Toward ecocomposition.* New York, NY: SUNY Press.

Ecker, P. S. (1995). Why define technical communication at all? *Technical Communication: Journal of the Society for Technical Communication, 42*(4), 570–571.

Fiksel, J. (2003, December). Designing resilient, sustainable systems. *Environmental Science and Technology, 37*(23), 5330–5340.

Ivanko, J. (2008, June). Ecopreneur or entrepreneur: What's the difference? *Sustainablog.* Retrieved from http://sustainablog.org/2008/06/ecopreneur-or-entrepreneur-whats-the-difference/

Jones, D. (1995). A question of identity. *Technical Communication, 42*(4), 567–569.

Langstraat, L., Palmquist, M., & Kiefer, K. (2010). Restorying disciplinary relationships: The development of an undergraduate writing concentration. In G. Giberson & T. Moriarty (Eds.), *What we are becoming: Developments in undergraduate writing majors* (pp. 50–66). Logan, UT: Utah State University Press.

Lowe, K., & Macauley, W. (2010). Between the idea and the reality . . . falls the shadow: The promise and peril of a small college writing major. In G. Giberson & T. Moriarty (Eds.), *What we are becoming: Developments in undergraduate writing majors* (pp. 81–97). Logan, UT: Utah State University Press.

Newcomb, M. (2012). Sustainability as a design principle for composition: Situational creativity as a habit of mind. *College Composition and Communication, 63*, 593–615.

Shamoon, L., Schwegler, R. A., Trimbur, J., & Bizzell, P. (1995). New rhetoric courses in writing programs: A report from a conference for New England writing administrators. *Writing Program Administration, 18*(3), 7–25.

Warnick, C., Cooney, E., & Lackey, S. (2010). Beyond the budget: Sustainability and writing studios. *Journal of Basic Writing (CUNY), 29*(2), 74–96.

CHAPTER 3

Reading University Ecosystems: Bolstering Sustainability and Revising Growth for Technical Communication Programs

Colleen A. Reilly

To thrive within contexts of scarcity, faculty who direct or teach in technical communication programs can benefit from embracing tenets promoted by movements that employ systems thinking, such as ecocriticism and sustainability studies. As Wals and Corcoran (2012) explain, sustainability studies promote anticipatory thinking, "seeing relationships and interdependencies," and altering values through cultivating "empathetic understanding and open mindedness" (p. 24). Systems thinking is especially important for technical communication programs that are small, recently developed, and/or embedded within larger interdisciplinary departments.

This chapter explores how faculty in technical communication programs can learn to read their university contexts as socio-ecological systems (SESs) and determine, through developing systems literacies, how to increase the sustainability of their programs through interconnectivity and integration within multiple levels of their ecosystems. My exploration begins by reviewing some of the theoretical bases for reading SESs, like universities, as ecosystems. I then outline some of the structural and cultural barriers that faculty face when attempting to pursue interconnectedness across traditional structural boundaries within universities and use examples from our program's experiences to illustrate how to reduce barriers and pursue integration at various levels. I conclude with a reminder of the need to accept uncertainty in these "post-normal times" (Wals & Corcoran, 2012, p. 27), in which the future of our universities will evolve in ways that are radically different from the past.

UNIVERSITIES AS ECOSYSTEMS

What do university structures have in common with natural ecosystems? Scholarship related to socio-ecological systems (SESs) outlines numerous connections between complex human institutions, such as universities, and natural ecosystems, detailing the similarities that make it productive to use similar strategies to analyze and work within both contexts. For example, Wood argues that elements of human societies exhibit "the same patterns of systems connectivity, complexity, and nonlinear transformation that we observe in the organic world" (2012, p. 6). The importance of interconnectivity of all elements within the system stands out as a central theme within the literature about human and natural ecosystems. The systems approach highlights that no actor or element within a system is independent; all components affect each other, and what happens to one part is likely to impinge on all other parts—either directly or indirectly.

From a structural perspective, a campus ecosystem typically includes departments, colleges, the university, and in some cases, a multicampus system as well as all of the people and resources located within each part. To function productively within the university as ecosystem, faculty need to learn how to think beyond typical boundaries of the program or department and understand how changes, crises, or initiatives in other seemingly remote parts of the university affect both the whole and all other parts, including the technical communication program. Furthermore, as Cote and Nightingale (2012) argue, when approaching SESs, such as universities, actors need to look beyond structures and investigate the political and cultural configurations that support these structural configurations. Focusing on structural elements is insufficient because those elements are often the direct result of political forces or cultural norms.

Approaching the university as an ecosystem requires that faculty develop new literacies. As Wood (2012) explains, "Systems literacy combines the study of social history and cultural discourses with a technical understanding of ecosystem processes" (p. 5). Social, cultural, and political discourse represent the forces that Cote and Nightingale (2012) mention as central to SESs and as codified in the policies, procedures, mission statements, initiatives, and other text-based entities that shape the structures within university ecosystems. Killingsworth (2005) highlights the role that texts play in shaping all sorts of ecosystems, natural and human. For example, our university system recently completed the development of a long-term strategic plan designed to guide the direction of all campuses for the next five years. Although this strategic plan outlines the goals for the system as a whole, rarely mentioning specific programs, systems literacy requires faculty to strategically read beyond the surface of the text and imagine the manner in which their adoption will impact local contexts.

While faculty may tend to see such texts as distant or impenetrable, through gaining systems literacy, they can find ways to work within these texts to shape their local contexts. In our case, the university system's new strategic plan sets a goal to become a national leader in student learning assessment. To respond to this, our professional writing faculty should regularly review our assessment processes and be prepared to supply useful and actionable data regarding our students' learning whenever it is requested. As Killingsworth's (2005) work demonstrates, technical communication faculty are uniquely qualified to develop systems literacy and to read and intervene in ecosystems and the texts that direct their constructs through our backgrounds in rhetorical analyses and information design.

Within research about SESs, a scarcity of resources constitutes another central theme, particularly because much of this research is recent and documents the situation within institutions after the economic downturn of 2008. Since that time, scholars have acknowledged that growth in educational institutions, as in most organizations, has been made less possible or desirable; instead, sustainability has become the goal. The definition of sustainability continues to be based on the work of the 1983 Brundtland Commission, formerly known as the World Commission on Environment and Development (WCED): "In defining our terms, we choose to adopt a well-known definition of sustainability, which is that of meeting the economic, social, and environmental needs of the present and the future generations" (Porter & Córdoba, 2009, p. 324; see also Bookhart, 2012). The key to this definition is the idea of meeting needs; achieving sustainability depends on identifying critical needs and determining how to address them within a particular ecosystem. Success in doing so often depends on strengthening interconnections and integration between the entity involved and the other portions of the ecosystem. Everett (2008) emphasizes that solving problems within higher education will "require interdisciplinary systems thinking, integrative 'knowledge transfer' skills and collaborative problem solving. It will also require the social sophistication, 'soft skills' and cultural literacy to cooperate with people of diverse backgrounds and conceptual frameworks" (p. 239).

Finally, both ecosystem-based analyses and sustainability theory emphasize the inevitability of change and the need to productively adapt to new situations (Atiti, 2012; Wood, 2012). Because change is nonlinear and ecosystems are multifaceted, approaches to sustainability require flexibility, a constant reappraisal of context, and multilevel and long-term perspectives. As Tucker, Cullen, Sinclair, and Wakeland (2005) argue, short-term decisions can often result in crises. Structuring programs from the perspective of sustainability and systems thinking necessitates careful planning accompanied by the ability to respond productively to the unexpected—a difficult balance of activities to perform well. The remainder of this chapter will explore how technical communication faculty can read their ecosystems, use the information they glean to increase the sustainability of their programs, rethink previous assumptions about growth, and prepare to adapt to an uncertain future.

INTERCONNECTIVITY, INTEGRATION, AND INCREASING SUSTAINABILITY

In this section, I discuss how programs can increase their sustainability through pursuing interconnectivity and achieving integration within multiple layers of their ecosystems. My discussion illustrates these processes through examples from our program's actions, strategies, and challenges. Increased interconnectivity and integration enhance sustainability by making the program an essential component of the various other parts of the campus ecosystem. The health of the program reflects and proves to be essential to the health of the ecosystem as a whole. As Bradbury (2003) argues, "A systems approach reminds us that organizational life is full of interdependencies; for example, a new strategic redesign requires all people to pull together in ways they perhaps did not previously. We neglect interdependencies at our peril" (p. 176). My discussion explores how technical communication faculty can foster interdependencies. Additionally, integration comprises a revisioned type of growth, one that does not rely on traditional definitions of growth within university contexts, for example, greater numbers of faculty or majors. Instead, growth can be measured in the degree of interconnectivity and integration, in the numbers of students in courses and quantity of other departments or units that the technical communication program directly impacts, and with whom we collaborate.

Brief Description of Our Program's Ecosystem

Throughout the discussion below, I use our program as an example, making it important to provide some information about our ecosystem. Our professional writing program is a track within the English major requiring 42 credit hours. We have between 80 and 100 majors in our track during any given academic year. We also have a certificate in professional writing that requires 21 credit hours, including an internship, and enrolls approximately 30–40 students. Students need a 3.0 GPA in their major to qualify to enroll in an internship, making the certificate open to our proficient students from a variety of majors; some of our professional writing majors pursue the certificate concurrently with their degrees. The Department of English offers two other major tracks: literary studies and English education. The track in literary studies also enrolls between 90 and 110 students while English education enrolls about 40 to 50 students. Our department is located within the College of Arts and Sciences (CAS), which is the largest college at our institution by far, containing 22 departments and awarding three quarters of the academic credits. Only education, business, and health and human services are outside of the CAS. As Anthony Atkins and I detail elsewhere, our professional writing program, started in 2001, has been located in an environment of scarce technological resources for its duration (Atkins & Reilly, 2009). We have only one computer classroom in our building, accommodating many upper-level writing courses. Our introductory course, ENG 204: Introduction to Professional Writing, is often taught online or in a computer classroom in another building, with older hardware and software resources.

Our program is housed within an institutional ecosystem that is best characterized as a public, southeastern master's-granting university, although we have doctoral programs in select departments. During the past academic year, our university enrolled 12,387 undergraduate students. Our student body is over 60% female, over 82% Caucasian, and over 82% from within our state. Our institution is located in a city of a little over 100,000 people. The top employers in our city are in the medical, pharmaceutical, educational, and power generation industries. Local government also employs a significant number of people. As a result, the opportunities for students to remain in the area and use their degrees in professional writing are relatively limited. Several of our graduates have found local jobs in organizations related to the above industries, while others have turned to freelance work in journalism or Web design or have pursued work in other fields. Many others have chosen to leave the immediate area to pursue employment.

All of the elements of our ecosystem described above determine in some measure the program we can sustain, the students we can attract, and the ways in which we can and should direct our curriculum. For instance, because our institution is located in a geographical area that supports a relatively large number of local publications, many of our students work as writing interns and as a result, gain experience and interest in pursuing this sort of work, which may be responsible for the popularity of our minor in journalism. In contrast, our lack of technology-based industries, such as software development firms, makes focusing a significant portion of our curriculum on communication in this industry less pressing. Likewise, the scarcity of technological resources in our immediate programmatic ecosystem and the lack of high-tech majors, such as engineering, at our institution restrict the sorts of faculty we can attract and responsibly hire into our program. New faculty who focus on engineering communications or need intensive computing resources for their research would have difficulty thriving within our ecosystem regardless of their ability to successfully pursue interconnectivity and integration.

Decreasing Barriers to Interconnectivity and Integration

Prior to integration into other components of the ecosystem, barriers—structural and cultural/political—need to be identified and addressed. Structurally, as mentioned above, university ecosystems are composed of recognizable components, including programs, departments, and colleges/schools. With some notable exceptions, most undergraduate programs on our campus and, I suspect, on many others, are contained within one departmental unit. We do have a number of interdisciplinary minors on our campus, including a minor in information technology; but these initiatives enroll relatively few students. Most majors, with the exception of our new international studies major, are contained within one department, and the great majority of their required courses come from within their home departments. Historically, on our campus and on many others, integration between departmental units has not been the norm. We speak frequently about the importance of interdisciplinarity, but the fundamental structures of the university and the cultural and political norms that support them prevent its implementation on a wider scale. For example, funding, faculty teaching loads, and other resources are funneled through departments. When I teach a non-English course in an interdisciplinary minor, such as the science, humanities, and society minor, I have had to do so as an overload; courses not labeled as English courses cannot generally be part of my teaching load. These common structural boundaries prevent wider interconnectivity and integration across units. Such realities also give rise to political and cultural barriers that further reify the organizational structures. Programs and departments have traditionally competed, whether implicitly or overtly, for majors and the resources that accompany them. To control their boundaries, they often institute structural barriers through the curriculum that are supported by cultural and philosophical norms. These barriers include requiring prerequisites for enrollment in all upper-division courses and restricting the enrollment in many upper-division courses to majors, which further hamper the integration of programs and departments into the larger ecosystem and enhance the silo mentality that has been much lamented but is generally undiminished.

Reducing these types of barriers—structural and cultural—requires innovation, which, from the perspective of Hailey, Cox, and Loader (2010) involves the development of creative solutions within constraints. Systems literacy requires that faculty who are attempting to attract students from other majors develop an awareness of the demands of those majors in terms of hours and remove barriers that are not absolutely necessary. In the case of our professional writing program, we decided to diminish barriers to our integration into the English department and wider university by removing prerequisites from all but two of our courses, easily enabling students from outside of our major to enroll. Our course, ENG 313: Writing about Science, for example, no longer has a prerequisite, increasing access for science majors. At our institution, most science degrees require a greater number of credits: a BS in biology requires 70 credit hours and a BS in marine biology requires 75; as a result, science majors cannot take many electives, making upper-level courses in other departments that have prerequisites out of bounds. In most semesters since we removed the prerequisite, course enrollment in ENG 313 consists of 30%–50% of students from the sciences, which enriches the educational experiences of our majors, exposing them to the research practices and ideas of their peers from the sciences while offering students from the sciences an opportunity to improve their writing.

Making this sort of change requires a structural shift, altering the curriculum but, and perhaps more importantly, necessitates a cultural and political change. We have made a decision not to protect the course for our majors by opening it to all. Additionally, enrolling students from across the university with no restrictions makes the course composition unpredictable, which in turn causes the course to be more difficult to teach because successfully reaching a diverse population can be pedagogically challenging. Finally, students in the course are unevenly prepared. Faculty often express dismay at having to teach some students the information or skills that they should have learned in lower-level courses. As a result, the lower-level courses are designated as prerequisites, although the skills taught in them may be useful but not critical to completing the upper-division course and might be easily taught with a minimum of effort to the students in need of them. Thus, several cultural shifts are required to remove barriers like prerequisites and accommodate students who are less prepared in some ways but who contribute significantly to the course in others.

While opening courses to students from outside a program requires one sort of shift that can increase the program's integration within the campus ecosystem, determining what courses count for a major requires a slightly different cultural shift. Programs and departments often define their majors narrowly, including mainly their own courses as core requirements. While courses from other departments might be prerequisites or suggested enhancements, they are less often central to the degree. When we started our program, we did not have this luxury because we lacked the resources to deliver a complete degree program on our own. As my colleague Diana Ashe and I detail in a chapter about the initial development of our professional writing program, during the first four years we had only two dedicated faculty with expertise in professional writing (Ashe & Reilly, 2010). Because our number of undergraduate majors grew very quickly to approximately 100, we were forced from the beginning to rely on resources and personnel from outside our subdiscipline and even our department to develop a workable and robust program for our students. From the start, we integrated the professional writing program into the department and the College of Arts and Sciences (CAS) by incorporating many courses not necessarily considered to be integral to a professional writing program, such as courses in literature, creative writing, and journalism. Including these courses required us to have a more fluid notion of what a program in professional writing could include. Although we now have five additional faculty from our field in our program, we still maintain courses from these other areas to fulfill requirements for our major because this sort of interconnectivity and integration provides flexibility for students and allows them to develop a breadth of knowledge.

Other barriers to the integration of the program, its faculty, and students into the larger university ecosystem may not be as easily addressed at a programmatic level because they are codified structural realities that require changes that faculty cannot affect on their own. The structures, including departments, that organize the university ecosystem are inherent barriers to interconnectivity and integration by making it difficult for faculty to teach outside their departments as part of their normal loads; restricting the ability of units to share resources, including infrastructure like classrooms; and focusing faculty's time and energies on their own departments and schools, thus inhibiting their abilities to know about the work of and make connections with faculty in other units.

Since the start of the strategic planning initiative within our university system, discussion about the duplication of programs and other initiatives across the systems has arisen. Duplications also exist within campus ecosystems; these duplications of efforts, resources, and classes, for instance, negatively impact the sustainability of units and the campus as a whole. Despite the detrimental effects of these barriers to integration, the ecosystem's structures prevent their removal. For example, we cannot schedule our classes related to writing and technology in computer classrooms owned by other departments when they are not being used, even though they have hardware and software that we lack, because those classrooms and their resources are restricted to the faculty and students of the departments who own them. On our campus, departments are largely responsible for purchasing the technologies for their classrooms, a reality that encourages exclusive ownership of rooms and discourages the sharing of resources. Departments have to use their operating budgets for this purpose even though all students pay a technology fee. This policy was changed. These structural realities at our university discourage sharing and reduce integration across the ecosystem. Wealthier departments have little incentive to agitate for such changes as they have sufficient resources and may be unwilling to risk political capital to complain on the behalf of less fortunate departments or units. The difficulty in making these changes, however, should not discourage faculty from working together to enact them because, as the discussion below demonstrates, the benefits can be great, and joint efforts are necessary for change to occur.

Interconnectivity and Integration as the New Growth

Diminishing or removing barriers clears the way to pursuing interconnectivity and integration across the campus ecosystem. Such integration can be a new way of viewing growth and can help the technical communication program share resources with other portions of the ecosystem and achieve greater sustainability as a result. In *The End of Growth: Adapting to Our New Economic Reality*, Richard Heinberg (2011) argues that there will be no lasting recovery in our postcollapse future; we have reached the end of growth in the global economy and need to create a livable new normal. Heinberg asserts that worldwide, all new growth will be relative growth: "From now on, only *relative growth* is possible: the global economy is playing a zero-sum game, with an ever-shrinking pot to be divided among the winners" (p. 2). Heinberg's thesis certainly applies to the current situation in public higher education across the United States. Public higher education has experienced direct budget cuts and competition for existing resources that make growth as it has been defined in the past very difficult. Few departments on our campus are receiving any expansion positions of faculty; in fact, replacing retiring or departing faculty is not guaranteed. With smaller numbers of dedicated faculty, programs and departments cannot easily add significant numbers of new majors. As mentioned above, across our university system, most of the talk is of reduction not expansion. System leaders are debating ideas such as closing campuses and reducing duplications of programs. Although growth in any previous sense may not be possible, we can still work to increase the numbers of students in our courses and simultaneously bolster the sustainability of our major by ensuring that we can continue to offer a full range of courses important for our majors. Rethinking growth as increased interconnectivity and integration has helped us to adjust to the current scarcities. We have integrated our program into the requirements for and elective options of entities at all levels of our campus ecosystem, from the department to the college to our university studies program, which has increased our sustainability.

Technical communication programs, especially those in departments with other major tracks, like literary studies, should determine how they might increase their integration with other areas in the department while maintaining a solid program for their majors. This can be accomplished by encouraging students from other areas to take technical communication courses, particularly the introductory course, and by involving faculty in advising students from all disciplinary tracks within the department. Such initiatives enhance the cohesion of the department, provide a common experience for all majors, and make the best use of departmental resources. In our department, we also made an agreement with our colleagues in literary studies to integrate the tracks in our major in professional writing and literary studies by requiring the introductory class in each of all students in both. This move prepared students in professional writing to take upper-level classes in literary studies and made it more palatable for our colleagues in literary studies to accept a one-course reduction in the number of literary studies courses required for our students. This effort at integration also aided in the sustainability of our program by enrolling more students from literary studies in our introductory courses, allowing them to more easily go on to pursue our 21-hour certificate in professional writing. While we have not gained significant numbers of majors, we have increased the numbers of students in many classes, which aids in the stability of our offerings by allowing our classes to fully enroll and even requiring us to add sections. For example, whereas we previously offered only four sections of our introduction to professional writing course as of four years ago, we now offer six each semester, all of which fill. Great numbers of sections make it easier for our students to create workable schedules and progress through the program more efficiently.

Technical communication programs should also examine the next level of the ecosystem, which in most cases constitutes the college and the departments that compose it, to locate other opportunities for interconnectivity and integration. For us, the next sphere of our ecosystem is the College of Arts and Sciences. Collaborations with other departments within the college provide an avenue for ecosystem integration. For example, three years into the development of our program, we supported the creation of a minor in journalism that would be shared between our department and Communication Studies. This minor drew from and resulted in the creation of a number of new courses in our program, such as Professional Magazine Writing, and increased dialogue between our departments. Initially, this was not a completely comfortable fit as none of our current faculty in professional writing came from programs that included journalism, which is a distinct field with its own conferences, journals, and theoretical approaches. However, our journalism minor fills a gap in the university's curricula; we do not have a separate journalism department, and our Communication Studies Department did not previously have a formal track in journalism. Therefore, integrating journalism added to our sustainability by necessitating the addition of and support for course offerings from which all of our students can benefit and increasing the numbers of students in our courses because five of our upper-level professional writing classes serve as electives. Furthermore, the journalism minor has allowed us to argue for additional lecturers and one tenure-line faculty member to support the program. As a result, we successfully hired a PhD in journalism, which was initially difficult because hiring a well-qualified candidate from a completely different field requires significant research and preparation. Pursuing this hire allowed us to improve the quality of our students' experiences in the journalism minor while bringing more students in contact with our professional writing program. As Bottery (2012) argues, organizations that are successful are flexible, adaptive, and creative: "Now the argument for creativity and invention is strengthened by the earlier recognition that prediction and control in complex systems is highly problematic" (p. 461). In the case of integrating journalism into our program, we needed to increase the flexibility in our hiring process by learning to hire on a completely different timetable and following a new process. This hire and the minor prompted us to broaden our definition of professional writing as a field of study, particularly at the undergraduate level. We agree with Wals and Corcoran (2012) that sustainability problems result from intransigence in terms of values "coupled with a lack of imagination and creativity" (p. 26). As difficult as it is to re-vision disciplinary boundaries, in our case expanding our idea of what constitutes a viable program in professional writing helped us to improve our likelihood of sustainability.

We have also become involved in other transdisciplinary initiatives by offering our courses as electives for several other majors and minors within our college.

In fact, we make every effort to accept all offers to make our courses either requirements or electives for other programs. Many inquiries have come as a result of networking opportunities with faculty from other departments, which result from the involvement of our professional writing faculty in delivering presentations about writing instruction at the Center for Teaching Excellence and serving on various university committees. Through networking, the chair of Environmental Sciences learned that we offered several courses that might assist their students. These courses, including ENG 204: Introduction to Professional Writing, ENG 313: Writing about Science, and ENG 314: Writing and Technology, were included as electives for the BS program in Environmental Sciences; they are the only humanities or social sciences courses that count for this major. ENG 204, ENG 313, and ENG 319: Document Design are also included as electives in the information technology minor. Networking with faculty from across campus and arguing for the utility of our courses for their students allows us to strengthen our interconnectivity between our program and others within our ecosystem. This has resulted in increased sustainability; the courses listed above consistently have waiting lists, for example. Additionally, more students from outside of our major have enrolled in our certificate program.

Technical communication programs should also examine and create opportunities to integrate within the sphere of the wider university. This requires developing literacy in the university's general studies program, for instance, in order to weave technical communication courses into a program that reaches all students. On our campus, we have increased course enrollments by revisioning our courses, such as Writing about Science, to be part of the new University Studies requirements for writing-intensive courses. This change required us to articulate student learning outcomes for the course that are in line with those for writing-intensive courses in University Studies. Our SLOs make it clear that Writing about Science will teach students to critically analyze texts, conduct research to support arguments, and cite sources correctly. In order to gain approval for a course to be included in University Studies, faculty must become literate in the processes of university bureaucracy, understanding what paperwork to submit, how to complete it successfully, and when to submit it. While this requires additional effort, the interconnectivity that is gained is worthwhile. Even if only a handful of students take writing courses and find them worthwhile, several report this back to faculty in their home departments, which can result in those faculty recommending the courses to additional students. Interconnectivity is self-sustaining, providing unexpected and organically expanding benefits.

Interconnectivity and Access to Resources

In many public university systems across the United States, raises for faculty members have become small or nonexistent. In order to seek additional resources to support pedagogical initiatives, including the use of new technologies, faculty can seek partnerships with other campus units, including centers of teaching excellence. Anticipating where resources will be funneled requires that faculty develop literacy in campuswide initiatives so that they can position their programs and their research to benefit from these increasingly rare influxes of money and/or equipment. One type of campus-wide initiative that has become ubiquitous is the Quality Enhancement Plan (QEP). Many faculty ignore the QEP until they are required to learn of and participate in it. In contrast, on our campus, faculty members from professional writing have been involved from the start of our QEP process, serving on the planning committee and later the task force for implementation. The QEP project selected by our campus relates to expanding and improving applied learning. As campus leaders in applied learning, largely in the form of service learning, professional writing faculty have been recruited to work with the campus Center for Teaching Excellence to create and grow a faculty learning community around applied learning. For this work, two of our faculty have earned course releases. This time has allowed them to work on their own projects in exchange for focusing on applied learning during the semester of service, which was already part of their pedagogical approaches. Additionally, being early participants in the QEP and applied learning has enabled professional writing faculty to be in an excellent position to write successful internal grant proposals for salary support and other resources to support innovations in applied learning. As this example indicates, the current situation of scarcity makes it imperative for faculty to develop an awareness of the larger university ecosystem and the work supported at the campus level in order to position themselves and their programs to obtain the few resources that are available. Internal focus is no longer effective; reading and responding to the conditions at the larger ecosystem level is essential.

REJECTING POSITIVISM AND ACCEPTING THE CONSTANCY OF CHANGE

While the sections above emphasize how planning and concerted activities can assist technical communication faculty in improving their sustainability through integrating their programs within all appropriate layers of their campus ecosystems, faculty who approach their universities as ecosystems should recognize that planning for the future is not always possible. SESs, like natural ecosystems, can change suddenly and in novel ways. In addition to prompting an investigation of integration into the larger context, systems thinking cautions against believing that any view of an ecosystem, human or natural, is wholly accurate or complete (Bottery, 2012, p. 454). Wood (2012) argues that sustainability studies assume that subjects are complex and unpredictable (p. 14) and that "sustainability . . . necessitates a humble acknowledgement of the limits of human knowledge" (p. 3). Rejecting positivism and accepting the inability to anticipate all future change may help program leaders to develop the appropriate attitudes for working within current campus ecosystems.

As Porter and Córdoba (2009) explain, there is no formula for developing lasting sustainability within an organization; a willingness to remain flexible and aware in order to deal with the unknown is essential: "there is an ongoing process of examination, learning, reframing, and action, based on mutually agreed upon definitions of sustainability in a particular context" (p. 337). We are faced with the challenge of rethinking how to maintain our program's sustainability within changing conditions; this work is never complete, and no approach is without compromise. Tauritz (2012) identifies two competences necessary for successfully approaching uncertainty:

> The two competences in this category comprise a close weave between skills and attitudes. Letting go of the need for certainty, and having the ability to adapt to change and generate new knowledge are central to learning to tolerate uncertainty. A second competence that deserves to be mentioned is the ability to change personal beliefs which may facilitate a positive attitude regarding an uncertain situation. (p. 306)

Pursuing interconnectivity throughout the campus ecosystem can be one hedge against an uncertain future for the technical communication program. If the program has made itself an integral part of many other structures within its ecosystem, it is difficult to disentangle from the larger structure. Greater numbers of students who depend on the courses solidify the program's position within the ecosystem. Additionally, as described above, integration can prompt faculty to soften their conceptual barriers and resist narrowly defining their field to their own detriment.

Writing faculty have historically seen their place within the university ecosystems as being located within English departments; however, over the past decade, this has begun to change. On our campus, we have had conversations with faculty in Communication Studies about the synergies between our disciplines, especially as technologies play larger roles in both our fields. Writing and communication studies separated long ago within American academia; throughout the 20th century, these areas were generally located in separate departments. Current waves of change across the university, coupled with the depletion of resources of all sorts, may provoke a reconsideration of such boundaries. Our personal beliefs about our field, especially from a historical perspective, should not restrict us from locating more sustainable homes for our courses and students within our campus ecosystems. Such relocations could better prepare us for the externally motivated restructuring of campus ecosystems that are certainly coming.

CONCLUSION

As this chapter has argued, the austere conditions that many of us face in public higher education are most likely not temporary situations; they are conditions that will persist and within which we must learn to thrive as individuals, programs, and institutions. In order to continue to offer the quality programs in professional and technical communication that we have developed, we need to think creatively about sustainability. Systems thinking can prompt us to learn from natural environments how to read the larger context and respond productively. Andreadis (2009) explains,

> Two central characteristics of all living systems are their purposeful behavior and adaptability to changing environments for the purpose of survival. It is this latter feature that stimulates the organism to gather data from its surroundings, reset its goals, alter its form within certain physical limits, and modify its behavior. (p. 6)

We, like the living systems we observe and learn from, can be ready to alter some of our goals and behavior to best exploit our context. Some of us respond by reaching out to other disciplines and forging joint initiatives, while others create partnerships across institutions. Whatever the solutions, they will most certainly involve innovative thinking about obtaining resources and recognizing that the past may not be a good indication of future conditions. Embracing these challenges will allow faculty in professional and technical writing to proactively shape the future contexts for their programs rather than allowing their programs and working conditions to be shaped by them.

REFERENCES

Andreadis, N. (2009). Learning and organizational effectiveness: A systems perspective. *Performance Improvement, 48*(1), 5–11.

Ashe, D., & Reilly, C. A. (2010). Smart growth of professional writing programs: Controlling sprawl in departmental landscapes. In D. Franke, A. Reid, & A. DiRenzo (Eds.), *Design discourse: Composing and revising programs in professional and technical writing* (pp. 89–111). Fort Collins, CO: WAC Clearinghouse/Parlor Press.

Atiti, A. B. (2012). Exploring possibilities of organisational learning-based change and transition towards sustainability. In A. E. J. Wals & P. Blaze Corcoran (Eds.), *Learning for sustainability in times of accelerating change* (pp. 439–456). Wageningen, The Netherlands: Wageningen Academic.

Atkins, A., & Reilly, C. A. (2009). Stifling innovation: The impact of resource-poor techno-ecologies on student motivation and proficiency. In D. N. DeVoss, H. A. McKee, & R. Selfe (Eds.), *Technological ecologies and sustainability: Methods, modes, and assessment* (pp. 1–26). Logan, UT: Computers and Composition Digital Press/Utah State University Press. Retrieved from http://ccdigitalpress.org/ebooks-and-projects/tes

Bookhart, D. (2012). Sustainability: Shifting definitions and evolving meanings. In J. Martin, J. E. Samels, & Associates (Eds.), *The sustainable university: Green goals and new challenges for higher education leaders* (pp. 83–92). Baltimore, MD: Johns Hopkins University Press.

Bottery, M. (2012). Leadership, the logic of sufficiency and the sustainability of education. *Educational Management Administration and Leadership, 40*(4), 449–463.

Bradbury, H. (2003). Sustaining inner and outer worlds: A whole-systems approach to developing sustainable business practices in management. *Journal of Management Education, 27*(2), 172–187.

Cote, M., & Nightingale, A. J. (2012). Resilience thinking meets social theory: Situating social change in socio-ecological systems (SES) research. *Progress in Human Geography, 36*(4), 475–489.

Everett, J. (2008). Sustainability in higher education: Implications for the disciplines. *Theory and Research in Education, 6*(2), 237–251.

Hailey, D., Cox, M., & Loader, E. (2010). Relationship between innovation and professional communication in the "creative" economy. *Journal of Technical Writing and Communication, 40*(2), 125–141.

Heinberg, R. (2011). *The end of growth: Adapting to our new economic reality*. Gabriola Island, British Columbia, Canada: New Society.

Killingsworth, M. J. (2005). From environmental rhetoric to ecocomposition and ecopoetics: Finding a place for professional communication. *Technical Communication Quarterly, 14*(4), 359–373.

Porter, T., & Córdoba, J. (2009). Three views of systems theories and their implications for sustainability. *Journal of Management Education, 33*(3), 323–347.

Tauritz, R. L. (2012). How to handle uncertainty: Learning and teaching in times of accelerating change. In A. E. J. Wals & P. B. Corcoran (Eds.), *Learning for sustainability in times of accelerating change* (pp. 299–313). Wageningen, The Netherlands: Wageningen Academic.

Tucker, J. S., Cullen, J. C., Sinclair, R. R., & Wakeland, W. W. (2005). Dynamic systems and organizational decision-making processes in nonprofits. *Journal of Applied Behavioral Science, 41*(4), 482–501.

Wals, A. E. J., & Corcoran, P. B. (2012). Re-orienting, re-connecting and re-imagining: Learning-based responses to the challenge of (un)sustainability. In A. E. J. Wals & P. B. Corcoran (Eds.), *Learning for sustainability in times of accelerating change* (pp. 21–32). Wageningen, The Netherlands: Wageningen Academic.

Wood, G. D. (2012). What is sustainability studies? *American Literary History, 24*(1), 1–15.

CHAPTER 4

Take a Leap of Faith and Hit the Gym: The Impact of Austerity on Professional Writing at a Private College

Madeline Yonker and Michael J. Zerbe

The age of austerity has had an undeniable impact across the nation on higher education generally and professional writing programs specifically. "Doing more with less" and other such mantras have permeated higher education, from small colleges to large universities, and from the humanities to the arts to the social and natural sciences and engineering. Professional writing programs have attempted to adjust to the constraints imposed by austerity in many different ways, all with the goal—even if not always attainable—of maintaining high-quality teaching and currency with rapidly changing technology, which speeds ever onward no matter the financial situation, and conducting cutting-edge, useful research. And although austerity may be in fact meeting the goals of its proponents—let us not for a moment think that austerity is universally supported—its short-term savings in terms of money do in fact force a trade-off—a price. This price can take the form of time as well as money or other resources. In any case, austerity leaves professional writing programs scrambling: in a bittersweet ironic twist, professional writing programs are growing nationally even as austerity inflicts its pain.

Although not directly impacted by cuts to state higher education allocations, writing programs housed at private colleges and universities have, like their state-supported brethren, felt the sting of austerity. Our chapter will focus on three ways that austerity has impacted the undergraduate professional writing major and minor at York College of Pennsylvania, a private, comprehensive institution of about 4,600 students which offers programs in both the liberal arts and pre-professional fields. First, we will examine the decline of enrollment in the professional writing major. Started in 2002, the major grew rapidly to 100 students by 2010 but fell to 75 students by 2013. This decline mirrored that of a decrease in the total number of students enrolled at York College, and we will discuss some of the austerity-induced pressures on enrollment at institutions such as ours, including the inclination of students to major in fields in which employment prospects are most promising. Second, we will explore the impact of delaying technology (both hardware and software) upgrades on faculty and students in the professional writing program. Clearly, professional writing majors and minors, as well as faculty, need a solid grounding in digital communication to be successful in the 21st century. Technology outdated as a result of austerity makes it difficult to fulfill this vital component of our mission. Third, we will investigate impediments to faculty-industry collaboration that have resulted from austerity. To teach most effectively, faculty who teach in professional writing programs, especially technology-heavy courses, must keep up to date with industry practice and trends with respect to genre and usability. Austerity complicates the ability to establish such relationships. Professional writing programs at any kind of institution cannot reach their full potential given the limits levied as a result of austerity.

CHALLENGES IN CHOOSING A MAJOR

Enrollment at many private colleges and universities throughout the United States has remained flat or decreased, and a primary reason for this situation is "price sensitivity," although an overall drop in the number of traditionally aged college students may also be a factor (Kiley, 2013). Private colleges and universities are attempting to stem the flow of students away from their institutions by offering ever-greater discounts off of the advertised (i.e., full) tuition. (Only a small percentage of students who attend private institutions pay full tuition; most students receive financial aid in the form of grants, loans, or both.) According to an *Inside Higher Education* article that describes research conducted by the National Association of College and University Business Officers, the vast majority of full-time, first-year students who started at private institutions in the fall of 2012—86.9%—did not pay full tuition; in fact, the average discount for these students was 45% of total tuition plus fees (Kiley, 2013). Although tuition at York College of Pennsylvania is approximately half that of many other private institutions in the mid-Atlantic, the College has indeed been a part of the trend to offer more students more aid in an effort to maintain enrollment and continue to keep the opportunity of higher education affordable even in difficult economic circumstances.

Because York College is so much less expensive than other private institutions, we were not strongly affected by the recession and subsequent austerity at first. Indeed, from 2007 until 2011, our numbers of applicants and incoming first-year students remained close to their pre-recession levels as these students and their families looked to York College as a more affordable alternative than some of the other colleges and universities in which they were interested. We could not stay ahead of the difficult economy forever, though. Overall enrollment dropped by 2.1% from 2011 to 2012. As a result, departmental budgets at York College have been flat, some capital projects and faculty/administration hires have been postponed, and raises for faculty, administration, and staff have been reduced substantially.

Internal research at York College demonstrates that the reason for the decrease in enrollment is financial. Office of Admissions staff here ask students who choose not to enroll at York or who transfer elsewhere if they plan to attend college elsewhere and, if so, which institution. However, the most common response to this question is that these students do not plan to attend any college at all because of limited financial resources.

Enrollment trends for the professional writing major at York College of Pennsylvania basically parallel those of the institution as a whole. The professional writing major enjoyed a period of strong, sustained growth from its inception in 2002 until 2010, when we enrolled our 100th major, an occasion we celebrated with cake. Since then, however, the number of majors has dropped to 75, a 25% decrease. This decline has accompanied the 2.1% drop in the number of full-time undergraduates at York College as a whole; however, the 25% decrease for the major is more than 10 times greater than that for the entire institution. We attribute this difference primarily to the fact that students, perhaps with encouragement from their parents and high school guidance counselors, may be choosing pre-professional majors that are explicitly tied to careers that seem to have positions available for college graduates even in a difficult economy.

Clearly, fewer students are choosing professional writing as a major during this time of austerity. Although professional writing programs attempt to capture some of the career-focused allure of pre-professional majors that are so popular with students in this day and age (Zerbe & DelliCarpini, in press), or a "rhetoric of professionalism" (Weisser & Grobman, 2012, p. 41), and enjoy excellent graduate placement (see Weisser & Grobman, 2012, pp. 45–48 for an example), students and/or their parents may not be completely convinced that the degree will lead to gainful, stable employment. Professional writing programs are, after all, often housed in that bastion of the liberal arts (and thus perceived rightly or wrongly as a road to nowhere in terms of a career), the Department of English. Although professional writing programs represent a clear attempt to link a part of English studies to fields and career paths outside academia, we cannot point to careers as clearly defined as nursing, accounting, education, or engineering for our graduates, despite the fact that even some of these fields, especially education, have suffered losses during the recession of the past 6 years (see, for example, Looney & Greenstone, 2012). We can point to graduates who have obtained well-paying, secure positions as writers, editors, and social media specialists as well as positions in other fields such as library science and education, but some prospective students or those who are already enrolled at the College but undeclared may not be persuaded, even if they really enjoy and have a strong interest in writing. Consider this May 2013 email that we received from a high school senior who had been accepted and planned to start his bachelor's degree studies at York College that August:

> Dr. Zerbe,
> My name is David [last name deleted to preserve anonymity], and I am going to be attending York College in the Fall.
>
> I am contemplating Professional Writing as a Major and have not yet decided upon a career. I was wondering whether you could take some time to answer some questions about Professional Writing as a Major and its ability to prepare the student to be hired for and work in the professional workforce.
>
> Is Professional Writing a major that employers will value?
>
> Does Professional Writing prepare a student for today's professional workforce?
>
> Finally, is the focus achieved through the minor how the student markets their more specific expertise to employers?
>
> Thank you so much for your time, I look forward to and greatly appreciate any information you can give me.
>
> Have a great summer,
> David [last name deleted to preserve anonymity]

We responded to David and commented favorably on the value of a professional writing major to employers, on the ability of the major to prepare students, and the value of a minor, which we require professional writing majors to declare in an effort to channel their writing in a clear direction. My colleagues and I hope that David will be convinced.

Indeed, not all students are persuaded to take this leap of faith. Our sense is that we have made the professional writing program attractive enough to some of these students to declare professional writing as a minor, but they retain majors in, for example, business or biology. Given that York College is not a pure liberal arts institution—as a comprehensive college, we offer many highly popular pre-professional programs as well—professional writing cannot lay claim to as much pre-professional territory as it may inhabit at institutions with a stronger emphasis on the liberal arts, and thus, students may elect to major in what they see as a safer pre-professional subject while maintaining professional writing as a minor. It is also true that some students who recognize that good communication skills are valuable in any field see the minor (or, in the case of a few students, a second major) as a way to enhance their pre-professional credibility: far from viewing professional writing as a liberal arts component of their education, then, these students embrace the "professional" in the professional writing program.

Of course, not all professional writing programs are housed in English or broader humanities departments. As the 2010 *TCQ* special issue on Positioning Programs in Professional and Technical Communication makes clear, a substantial number of writing majors are also found in communication departments or as independent entities (Yeats & Thompson, 2010, pp. 233–247). Myriad reasons justify these organizational decisions, but one rationale might be that the programs can be perceived as more professional if they are not closely associated with traditional English departments.

Part of the problem, at least from the point of view of students, may be that fields like professional writing are not credentialed beyond the bachelor's degree at the state or federal levels. Students who pursue careers in education, nursing, medicine, law, and accounting, for example, must take state or national examinations and achieve a minimum score in addition to earning a degree to be able to begin their professional careers. Students in other majors such as engineering, sport management, or music do not take national exams, but the programs themselves are accredited by national organizations. Students may view these outside-of-academia requirements and associations, for better or worse, as evidence of a more well-defined field of study that provides opportunities for networking and a more proven path to employment. Those of us involved in the teaching and profession of writing have, of course, had many a conversation about the credentialing issue. For example, in *Editing Fact and Fiction*, Sharpe and Gunther (1994) point out the following:

> There are no postgraduate degrees in editing and, unlike professions such as law, medicine, or accounting, no qualifying examinations. Indeed, editing is as "unprofessionalized" as it was in the days when publishing was a "gentlemanly" pursuit. By this we mean that at most houses, editors are simply not taught how to edit; it is just assumed that they know how or else will pick it up as they go along. (pp. 4–5)

This scenario may be perceived by students to be too uncertain in terms of being able to demonstrate their knowledge and skills. While it is certainly true that some employers require applicants to complete a copyediting exercise, demonstrate knowledge of particular software, or otherwise provide clear evidence of their abilities (see, for example, Saller, 2013), students may feel more confident about finding employment with a degree in a major that offers them an opportunity to earn an additional credential.

Many faculty, of course, are happy that writing programs are not accredited nationally: such conformity is anathema to many in the humanities. Even so, attempts to develop common policy and standardize learning outcomes are occurring: the Council on College Composition and Communication (4Cs) and the Council of Writing Program Administrators (CWPA) both have expended considerable energy and time in this regard. For example, 4Cs published "A Position Statement of Principles and Example Effective Practices for Online Writing Instruction (OWI)" in March 2013, while CWPA's "Outcomes Statement for First-Year Composition," first composed in 1999, has proven highly influential. At this point, although they will furnish writing programs with more of a pre-professional aura, they don't offer any kind of postgraduate credential and thus will not likely have any impact on employment prospects for our students. On the other hand, some colleges and universities (e.g., Drexel University, University of Texas at Arlington, University of California at Santa Cruz) have developed certificate programs in an attempt to fulfill this perceived need.

Fortunately, although the number of professional writing majors has decreased, the recession did not reduce the number of faculty in our professional writing program. In fact, we have enjoyed a net increase of two full-time, tenure-track faculty members (for a total of five) since 2008. The first was hired in 2008 while the major was still growing rapidly and the recession had not yet impacted York College, while the second was hired in 2013 for a faculty line designated largely for first-year writing as a part of general education reform but which will also play an active role in the professional writing major and minor. Even with fewer students in the major, we certainly needed the additional faculty: the professional writing major remains the largest major in the English and Humanities Department, of which we are a part. We must assume, though, that it is likely that the number of Professional Writing faculty will increase no further unless the major begins to grow again.

One way to address students' wariness about networking, credentialing, and career prospects associated with the professional writing major is to initiate career-related discussions from the very start of the major. We begin talking with our students about internships when they take our sophomore-level Writing in Professional Cultures course, and in an academic advising capacity, we have often quite extensive discussions with our students about their choice of minor. A minor is required for the professional writing major at York College; the goal is to have students channel their writing into a specific direction that they envision as part of their future lives as professionals. We hope that the early start on internship possibilities and the choice of minor impel our students to seriously consider their future plans and how their academic experiences will—without sacrificing the many advantages of a liberal arts education—enable them to realize their goals.

It is useful to place enrollment trends in professional writing majors into a broader context. From 2002 to 2010, the number of professional writing majors at York College grew swiftly and with little effort on our part in terms of recruiting other than identifying and speaking with good writers in our first-year composition courses. We were the cool new major, and we enjoyed a prolonged honeymoon period that took advantage of the shift in literacy practices from reading to writing as chronicled by Brandt (2009), who boldly asserts that "for perhaps the first time the history of mass literacy, writing seems to be eclipsing reading as the literate experience of consequence" (p. 161). (Also see Lee, 2007, p. 225, for example, for a discussion of texting as a literacy practice). By 2010, though, it was clear that the honeymoon was coming to an end, exacerbated at least in part by the enrollment decrease at the College. As a result, we began to initiate more systematic recruitment efforts. While we continued to use the fertile recruiting ground of first-year writing classes, we also produced a video featuring Professional Writing students and the coordinator of the program (Mike Zerbe) touting the major and large posters designed by two professional writing majors for high-visibility bulletin boards. It is too soon to determine if these efforts have had any effect, but we feel that in an age of austerity, we must work harder and smarter to ensure that our program enrolls enough students. For example, we may ask our writing tutors (who work in our Writing Center) and Writing Fellows (who are placed into first-year writing classes) to be on the lookout for good writers and to alert these students to the opportunities that the professional writing major or minor would offer them.

The increase in the numbers of professional writing majors and the numbers of students in these programs has occurred nationally (Balzhiser & McLeod, 2010, p. 416; Giberson & Moriarty, 2010, p. 2) even as the overall numbers of English majors has dropped (Miller & Jackson, 2007, p. 683). However, some intriguing signs point to a renewal of interest in English studies and other humanities disciplines. For example, research on companies and organizations that employ recent college graduates reveals that many employers value traits and characteristics such as strong communication skills, the ability to adapt to change quickly, and the ability to synthesize and integrate knowledge in their new employees that are more common in humanities and liberal arts programs than they are in pre-professional programs (Forsythe, 2012). (The lack of communication skills is not only a college-level problem: we have added a number of new sections of our developmental writing course over the past few years, which shows us that despite the widespread enthusiasm for and practice of perhaps a lot of informal writing, students may not be gaining the literacy skills they need because they may not read and write in an academic/professional context as much as they used to.) Also, in the United Kingdom, for the first time in decades if not centuries, new liberal arts programs are being developed and are enrolling students in an effort to, in the case of University College London, "bridge the chasm dividing" C. P. Snow's two distinct cultures of the humanities and the sciences (Guttenplan, 2013). Like most European universities, students enrolled in British postsecondary institutions study only one subject. According to Guttenplan (2013), "Students, who may take a dozen different subjects at age

14, are expected to filter down to just three or four by age 16 in preparation for applying to study a single subject at university." It appears that the tide may finally be turning away from narrow specialization and back to the broad-based education that is a hallmark of the humanities. The new General Education program at York College will attempt to take advantage of this shift: two semesters of writing/communication will still be required of all students, although the second course may now be housed within major programs, and a new required, interdisciplinary (for most students), communication-intensive first-year seminar will be added to the General Education curriculum, thus ensuring that students will have additional synthesis and communication experience.

THE COST OF A PROFESSIONAL WRITING MAJOR

One challenge the professional writing major at York College has faced, as is noted above, concerns the costs of writing technologies. And as is the case with any discipline whose technologies advance quickly, that challenge is twofold: The first is in finding the resources to purchase and update software and hardware for the students to use in the classroom; the second is in supporting faculty's professional development in order that they are able to properly instruct students on the use of such constantly changing technologies.

Early in the professional writing major's existence, it became clear that our costs would—or could—exceed our small departmental budget. Historically, specific departmental fees—beyond tuition and general student fees—were incurred only by students majoring in the natural and physical sciences and fine arts; now students in many more majors, including professional writing, according to a preliminary discussion here at York College, may have to pay these additional costs. These increased costs, if ultimately implemented (no final decision has been made), would mirror those of the increased expense of writing in society as a whole. No longer a refuge of the starving artist who could purchase pens and paper on the cheap, writing—at least writing that is meant to make an impact beyond the writer and perhaps a small circle of like-minded comrades in arms—needs expensive, sophisticated technology, equipment that can be difficult to afford during a period of austerity. At York College, the discussion has focused on lab fees for required courses for the professional writing major such as Digital Writing and Document Design, which use software and hardware resources heavily. And while the added fees (again, if ultimately imposed) would not increase the cost of a college education substantially, they would pose enough of a burden that some students would perhaps limit their options for majors to those that do not impose additional financial hardship.

However, so far the program has been fortunate to find the necessary resources to maintain two designated computer labs: one classroom for instruction and an additional lab for student use. Each lab has 20 PCs, which are loaded with

Microsoft Office along with the other standard Windows programs (Notepad, Paint, etc.). The machines are updated every 3 years. In 2009, we were able to add licenses for Adobe Creative Suite to the machines in each lab. This gave our students access to programs like Photoshop, Fireworks, Dreamweaver, and InDesign—many programs that writers who work in publishing and Web design use.

The addition of the Adobe programs allowed us to expose our students to two WYSIWIG Web editors (Fireworks and Dreamweaver) and to photo-editing via Photoshop; also, with the addition of the Creative Suite, we were able to develop a course in print document design. However, the addition of the Adobe programs also created challenges: the machines in ours labs were not robust enough to run the Adobe software efficiently; often start-up would take several minutes, and the machines would lock up occasionally, requiring a hard restart. Also, the versions of the software are significantly different on faculty machines (several of whom have Macs) than it is on the PCs in the labs, so faculty were often preparing lessons on the lab machines in order to maintain consistency in instruction protocols. Moreover, some faculty needed to learn the programs in order to use them in the classroom. Finally, the question of keeping the software licenses current presented yet another budget obstacle. As the college froze departmental budgets in 2010, the resources to maintain and upgrade both our hard- and software were decidedly scarce.

While these challenges are borne primarily out of austerity measures our college has felt campus-wide, they have been extraordinarily useful. Such challenges have forced our Professional Writing faculty to examine the value of teaching specific software tools as part of our larger curriculum. In other words, we were forced to answer the question, Should the program be responsible for teaching its professional writing majors how to use specific software? Furthermore, we have had to determine the value of our own professional development that would be in service of learning new or remaining current with particular programs, which is time-consuming and often undervalued by humanities departments and tenure and promotion committees.

AUSTERITY CREATES SPACES FOR RETHINKING THE PROFESSIONAL WRITING CURRICULUM

As we have witnessed in the jobs our own graduates have been placed in, and as scholars like Alex Reid (2010) note, the expectations for a professional writing curriculum are varied and ever-expanding (p. 254). Along with the shifting and expanding expectations for our graduates, we're also seeing the needs of our incoming students broaden. That is, the preparation our writing majors have before they enter our program is increasingly diverse. The effect, then, is one of several moving targets that our professional writing curriculum must hit: the changing student needs in basic writing skills as well as the seemingly constant addition of new skills and tool knowledge they are expected to have upon graduation. Addressing both the diverse writing needs of our students on the front end of their undergraduate education as well as the expanding needs of the industry presents two challenges for professional writing programs: curriculum development to best serve our students and professional development that best prepares our faculty.

Research on the curriculum of the emerging professional writing major agrees that the industry requires graduates to not only have training in writing theory/rhetoric, but also practical skills for professional writing situations and in visual design. David Beard (2010), writing about the development of the professional writing major at the University of Minnesota Duluth, describes their program as "the advanced study of writing in professional contexts [which is] complicated and enriched by reflection on and practice in generating texts within the diversity of print and visual cultures" (p. 188). Other skills that professional and technical writers can be expected to have include "printing, graphic design, publication management; information management; communications technologies; and often laboratory or production experiences in nonprint media, including video (Meese & Walhstrom, 1988, p. 33). Based on exit interviews and contact with our own alumni, we would add that our graduates are also best served when they have exposure to social media tools in a professional context and to the rhetoric of social media management.

Austerity forces the question of *how* we should go about preparing students for careers in print and Web design, technical writing and communication, and writing in environments like social media. Austerity forces the question of *which* tools we should give primacy to. The answers, as our program has fashioned it, relies on scholarship like that of Alex Reid (2010) and Stuart Selber (2004), which acknowledges the importance of training in technologies and multiliteracies for writers, but underscores the importance of a rhetorical, critical approach. Reid (2010) argues that professional writing's "disciplinary goal should not solely be how to integrate . . . specific [technologies] but also how to create curricular structures and practices that will allow us to deal on an ongoing basis with emerging technologies" (p. 255). In other words, our curricular focus should not be on the technologies themselves, as the tools our students encounter after graduation will most likely be changed or different by the time our students begin their careers. Instead, our attention should be on ways we and our students can best remain current. The argument can be made that even if the program were to have the most recent software for our students, the rate of obsolescence is so great that by the time they graduate and begin working, the versions are likely to be different or the industry standards for a particular program might be completely changed.

The most useful example of the ways York College's professional writing curriculum shifted to both address the lack of resources as well as acknowledge this inevitable march of obsolescence is in the Digital Writing class, a 300-level course meant to prepare professional writers for designing and maintaining websites, creating information hierarchies and navigation, and composing online content. Originally conceived as a Writing for the Web class primarily focused on familiarizing students with Dreamweaver, this was one course wherein the decision to remain current (or not) with Adobe Creative Suite would be crucial. Additionally, the faculty person originally teaching the course had taken a course in using Dreamweaver in order to teach students the software, so additional training for faculty would be necessary if the Professional Writing program were to continue to update the software.

In 2010, with the budgetary freeze in effect and the computers in the labs running slowly with the Adobe Creative Suite, the Writing Program Committee approved a revision of Writing for the Web to include a new title (Digital Writing) and revised curriculum with a shifted focus. While the original iteration of the course was grounded in rhetoric, with issues like audience analysis and user access clearly addressed, the revised course placed more emphasis on theoretical, contextual, and critical issues of technorhetoric. Moreover, the revised course included a rather involved assignment called the "New Tool Demo," in which students were responsible for seeking out a new (or unfamiliar to them) Web-based writing tool. Then they were tasked with determining the purpose of the tool or what the developers explicitly intended as the purpose of the program (i.e., real-time collaborative writing, easy and free photo editing, flash-based Web design, social media/networking, etc.). Next they were to teach themselves to build something, compose a document, or otherwise use/participate with/in the tool. Finally they were to develop a short protocol for basic use of the tool and to present their product to the class with a reflection on the affordances and limitations of the tool, usability advantages and disadvantages, and potential ways the tool might be hacked or otherwise used for user-generated purposes.

The revised Digital Writing course, therefore, shifted its focus from being a "tool-centered" course, in which the students learned to write in and with a specific writing technology, to being a course in which students were able to develop strategies for interacting with and teaching themselves an unfamiliar program, to reflect critically on how certain writing tools and media effect messages and writers, and to practice producing one fairly ubiquitous professional/technical writing genre: the protocol.

Furthermore, the revised Digital Writing course was no longer predicated upon teaching students how to use a specific version of a software program, which therefore meant the program no longer needed to be as concerned about keeping the Adobe Creative Suite licenses current. And while the course still offered students a brief introduction to the Adobe tools in the labs, the course outcomes did not assume that students would achieve fluency in one program; rather, the outcomes described writers' ability to develop methods for working in unfamiliar digital writing environments and to understand the ways that writing technologies act as agents in the construction of audience and message.

AUSTERITY OPENS OPPORTUNITY FOR PROFESSIONAL DEVELOPMENT FOR PROFESSIONAL WRITING FACULTY

A second way in which our Professional Writing faculty have managed to improve our curricula amidst shrinking resources is to revise the way we think about professional development. Attendant to the challenges our Professional Writing faculty have faced in finding resources to properly prepare our graduates for industry work, we have also faced the additional challenge of keeping ourselves current with concerns and trends in the industry itself. Again, as with so many disciplines whose fields are constantly changing as technology and research advance, the technical and professional writing industry present a particular challenge to Professional Writing faculty, especially those of us whose personal graduate training focused primarily on writing and rhetorical theory.

The material conditions and practical concerns that professional writers face are often overlooked in graduate programs whose emphasis are on rhetoric and composition. This became clear during the first semester that we ran the revised Digital Writing class: students dealt with clients who presented a host of pragmatic obstacles that the course struggled to support. For instance, students' clients had existing websites hosted by third parties that they wanted overhauled, but the third party was resistant to releasing information or offering our students access (to administrative passwords, for instance) or was unreachable. Students had clients who failed to deliver necessary images and content the student needed to complete the site. Students had clients who insisted on design and layout schemes that were professionally unacceptable and were poor for usability and intuitive user navigation. Students had clients who were resistant to the students' advice concerning the inclusion of social media as a potential use of the Internet to better reach out to their customer/audience base.

Such concerns and their possible solutions are not immediately found in the disciplinary scholarship; that is, these material issues of dealing with "real-world" problems that professional writers face with clients and stakeholders are not always addressed in research. Therefore, in order to be able to address such material concerns in the classroom, it is useful for Professional Writing faculty to have regular interaction and exposure to nonacademic professional and technical writing environments.

Research in both professional development and the field of rhetoric supports the argument that faculty can benefit from alternative models of professional development outside the conventional scholarship model. Such alternative models encourage conducting research outside our own specialty or programmatic discipline (Greenhow, Robelia, & Hughes, 2009); argue that professional development be holistic, authentic, hands-on, and experiential (Webster-Wright, 2009); and imagine professional development as an avenue for engagement wherein the faculty member becomes an ambassador for the institution to the larger community (Sandmann, Foster-Fishman, Lloyd, Rauhe, & Rosaen, 2000). Additionally, the field of rhetoric, whose theory underpins our curricular goals for the professional writing major at York College, has in the last decade acknowledged that rhetoricians should contribute to the "public face of rhetoric" through "participation in public discourse" and through consulting work (Zarefsky, 2004, p. 36).

Thus, in order to enrich our professional writing classrooms without the requiring additional funds, faculty are currently using professional development time for unconventional yet academically and pedagogically useful projects. Such projects are meant to fill gaps in faculty knowledge, provide faculty with examples and anecdotes for classroom teaching, build faculties' writing portfolios/résumés, and potentially result in additional research/publishing opportunities.

One such project involves interviewing local Web designers to determine on-the-ground strategies for interacting with clients. While clients have specialized knowledge of their own company, product, or service, often they do not acknowledge the specialized knowledge that the writer has: that of effectively reaching and communicating with a specific audience. Writers who work in digital writing often find that their clients have preferences for design and delivery that defy professional writing and design conventions—and defy those conventions to the detriment of their own website's ethos. Therefore, professional writers who work as Web designers and Web content developers often find themselves in the position of educating their clients (or disagreeing with their clients) about the rhetoric of online communication. Often the professional Web writer must justify decisions he has made in composing and designing the client's website—decisions that may not align with the client's preferences. While such interviews might not be considered scholarly, as they are not positioned as ethnographic or critically methodological, data from these interviews—essentially the strategies that Web designers leverage in order to both please clients as well as produce rhetorically effective websites—will be useful to our students.

Strategies that local Web designers and content writers report using can easily be brought into the classroom for students to consider and practice during their own client-based work. For instance, one Web writer reported that he preempts clients' preferences for potentially unprofessional design choices by preparing several mock-ups for the client to choose from very early on in the process—often for the very first consultation—so that the client does not feel responsible for making substantial contributions to aesthetic, navigation, and architecture decisions for the website. Another interviewee reported that during the early stages of the invention process, she provides her clients with a kind of "market analysis" comparison—examples of websites designed for similar purposes to her clients'—for her clients to look over and use as a starting point for decisions about menu items and layout. Such strategies, leveraged early in the design process, allowed these professional Web writers to both work collaboratively with clients as well as maintain significant control over design outcomes. The usefulness of such anecdotes is clear in the classroom; the addition of either strategy as a step in their client-based project protocol would both allow students to practice developing a market analysis for a client (or to render multiple mock-ups early on) as well as hold such strategies in reserve for when they must interact with authentic clients upon entering the professional world. Also, if this particular set of interviews were expanded via a broader pool of interviewees, patterns in strategies for productive collaboration with clients could be easily developed into a more rigorous, academic argument and reported in a published article. Therefore, such unconventional professional development proves to be significantly cost-efficient for the faculty, as it both enriches pedagogy as well as provides a potential site for more traditional-style academic scholarship.

Another such project involves one faculty member consulting as a writer/social media developer for a local business. While the position is unpaid, the consulting works as a kind of internship, in which the benefit to the faculty member is in industry experience that she can take back to the classroom and deliver to her students. In this case, the faculty member volunteered her services to the owner of a small local fitness center who faced competition from a new gym of the same brand affiliation that had opened across town. The owner was concerned about distinguishing his gym from the new one and about the potential loss of clientele who might find the other location more convenient.

In consultation with the owner, the faculty member proposed a layered approach to creating a distinction between the two gyms: the development of a mission statement for the owner's gym, the launching of a campaign to mark the owner's gym as one of the original gyms of this particular brand, and the production of a series of ongoing athlete profiles of the owner's current clients.

The implementation of this multifaceted project put the faculty member firmly outside her theoretical wheelhouse; as a digital writer, she'd studied digital rhetorics and networked communication and had developed several noncommercial websites. Moreover, as a client at the gym but not an expert in fitness, the faculty member had the additional challenge of appropriating the discourse of a specific industry.

The mission statement was produced after a thorough perusal of other affiliated gyms' mission statements as well as several discussions with the owner and the trainers at the gym. The legacy branding—the strategy to distinguish the owner's gym from the new as having more years of experience—was conducted through a series of posts on the website that pushed out to the gym's social media networks (Facebook, Twitter), which involved discussions of virtuosity and standards of movement for athletes. These posts required significant research on the part of the faculty member along with several collaborative drafting meetings with the owner to ensure the language and content were accurate. Finally, the development of the athlete profile, which also became posts on the website as well as links on the social media networks, built community for the owner's clients by showing the athletes he profiled that he thought them valuable enough to display their progress and insights, and by introducing individual clients to other clients and encouraging interaction among them in the comments on Facebook. Again, the production of these athlete profiles required the faculty member to interact with the athletes to develop the profiles and to take photos to post on the gym's Facebook page.

Working as a writer and in social networking for the gym required the faculty member to inhabit a space of professional unfamiliarity. She launched a professional social media campaign and developed a branding campaign, two activities she'd not been asked to do in graduate school. She had to reimagine the gym's professional website as a site for personal interaction with customers, which required her to teach herself how to use Drupal, a platform she'd not worked with previously. Also, the faculty member had to become fluent in an unfamiliar, specialized discourse in order to write with the proper ethos for the gym community. And while this particular example of professional development has less direct connections to the classroom, pedagogy, or to potentially more academic outlets, the faculty member can now bring the ethos of having worked as a social media coordinator to her students. She now has personal experience in the creative and critical processes of problem-solving for a business through writing and media. Further, she has insight into the challenges of writing in unfamiliar discourses for specific industries and can develop classroom activities to better help her own students work through similar challenges they will face in entering unchartered industry discourses.

AUSTERITY REVEALS ALTERNATIVES TO CAREER DEVELOPMENT AND PLACEMENT

There are other ways this model of professional development combats the struggling economy's effect on faculty and students. For instance, our students' concerns, discussed above, about the potential for employment with a degree in professional writing cannot be discounted. With traditional full-time, benefits-providing jobs scarce and coveted, graduating professional writing majors often face a daunting task in seeking employment that they feel justifies and makes use of their degree. Indeed, it can be argued that rhetoric is even more important during times of austerity than it is during those of prosperity. We and our students must make clear, effective arguments to have any chance of weathering the current economic storm successfully. Only the most skilled uses of rhetoric will result in positions and resources for writing majors both while they are in college and after they graduate and pursue employment. The strain of job placement in the struggling economic environment has forced us to implement a number of alternative strategies to help our professional writing seniors and recent graduates get a promising start after college. These alternative strategies draw significantly upon the connections that our faculty make in industry as they work as consultants and as they interview local professional writers.

Our first alternative strategy is to attempt to convince our students to go beyond Monster.com and other similar mass employment websites and to zero in on positions related to a professional writing major by becoming familiar with national and local professional organizations for writers. These organizations range from the Society for Technical Communication to the Public Relations Society of America to the International Food, Wine, and Travel Writers Association to the Society of Environmental Journalists. Websites for many of these and similar organizations have names of national and regional officers as well as contact information that students can use for networking. Many of the sites also have job boards, though some require a membership to join. When this is the case, we tell our students to join only if they are seriously interested—noting that membership fees are often far cheaper for students than for professionals—and to use the contact information for officers and others associated with the organization to begin networking in lieu of joining.

On a related note, we ask our students to think outside the box in terms of job titles. Professional writing majors do not go on to become only writers, editors, social media specialists, researchers, journalists, and public relations professionals; they also find rewarding careers in insurance, library science, retail management, event planning, teaching, and law. We have alumni who have pursued leads in all of these fields and have achieved success. Beyond job titles, we tell our students to think of themselves as not only writers and communicators but also as information managers and information architects. These titles help students to think of their abilities to organize and design information, skills that are especially important in this digital communication age.

Second, we tell our students that they should think of their professional writing major, including the required internship, as equivalent to several years of professional experience in writing, editing, and related fields. We encourage students to apply for any position in which they're interested that calls for 5 years or fewer of experience. Our students have found that some prospective employers are pleasantly surprised to discover that our program includes courses such as Digital Writing, Document Design, and Professional Editing, and employers have remarked that these courses provide students with experience that is equivalent to on-the-job training. For example, several employers have been delighted to learn that our students know copyediting symbols and terminology such as *stet.* Additionally, to help students navigate the complexities of the job market, we invite local and national writers and editors as guest speakers to campus with the intention of gathering insight and insider knowledge about how employers read cover letters and résumés when they're hiring writers. For instance, last year we hosted Maggie Koerth-Baker, the science editor for the online magazine *Boing Boing* and author of *Before the Lights Go Out: Conquering the Energy Crisis Before it Conquers Us*. She proved an invaluable resource to our majors, offering them everything from business plan templates to advice about how much to charge for freelance jobs. We also hosted Jason Konopinski, a local Web content developer, who discussed the way he was able to use his blog about writing and creativity to construct a network of readers and writers that eventually led to his landing a job with a large, local marketing software company.

Third, we are working on two methods of maintaining the network of current students, alumni, faculty, and local industry contacts. One of these in-house networking opportunities is the York College Professional Writing Students, Alumni, and Friends Facebook group. This group allows us to keep in contact with alumni who often post jobs and opportunities that our current students might take advantage of. Additionally, any leads faculty are given by local industry contacts are posted so that both current students and alumni can access. The second networking opportunity is one we are currently developing: an annual Professional Writing Alumni Panel. The panel will consist of alumni working in a range of positions, and we'll bring them together with current students for an informal gathering so that current students can understand the different opportunities and paths that their career can potentially take, as well as make face-to-face connections with people in a variety of industries who could potentially be useful contacts as the students find themselves in the job market.

Fourth and finally, we introduce our students to the business and practice of freelancing, which we have found that many of our seniors, knowing the dismal economy they face, ask about. This discussion often takes place in Senior Seminar or in Professional Editing, another senior-level course. We remind our students that they have, of course, already gained a great deal of freelance editing experience helping friends and classmates with their papers; additionally, many of our students have worked as writing tutors and/or Writing Fellows and thus procured additional experience. The task now facing our students is to take this experience outside of the college setting: we discuss marketing, estimates, contracts, billing, taxes, and other issues facing small-business owners.

CONCLUSION: THE BRIGHT SIDE OF AUSTERITY FOR YORK COLLEGE PROFESSIONAL WRITERS

At York College of Pennsylvania, the professional writing major has felt the pinch of the past decade's recession. The challenges that our program and programs across the country should not be downplayed: each dollar spent by the program is mightily earned by the faculties' efforts to seek out resources. Each semester, we must constantly argue for funds to pay honoraria to our guest speakers; we must ask favors of our local contacts who volunteer their time to visit our classes; we must write proposals to our chair and to our Academic Dean to ask for funds to cater gatherings; we spend additional time on maintaining personal local networks with industry writers; we spend additional time working ourselves as social media managers for our Facebook group; we spend additional time on consulting and industry work above and beyond the standard expectation for scholarship and professional development. Our teaching loads remain the same in spite of flat budgets and additional responsibilities.

However, it is impossible to avoid the obvious and clichéd Pollyanna approach that our program has taken with regard to the stringent economic environment. Necessity has required our faculty to experiment with and think critically about the content and skills we teach and our pedagogical methods in delivering the same. Those experiments have resulted in, we believe, improved classroom practices; more practical and pragmatic curricula; a more rich, diverse approach to conducting our own scholarship; and better preparation of our students for the job market.

ACKNOWLEDGMENTS

We gratefully acknowledge the assistance of Stephen Neitz, Assistant Dean for Enrollment Management, and Debra Staley, English & Humanities Department Administrative Assistant both of York College of Pennsylvania.

REFERENCES

Balzhiser, D., & McLeod, S. H. (2010). The undergraduate writing major: What is it? What should it be? *College Composition and Communication, 16*(3), 415–433.

Beard, D. (2010). Writing studies as grounds for professional writing: The major at the University of Minnesota Duluth. *Programmatic Perspectives, 2*(2), 181–189.

Brandt, D. (2009). *Literacy and learning: Reflections on writing, reading, and society.* San Francisco, CA: Jossey-Bass.

Forsythe, G. B. (2012, November 5). The value of a liberal arts education in today's job market. *Under 30 CEO.* Retrieved from http://under30ceo.com/the-value-of-a-liberal-arts-education-in-todays-job-market/

Giberson, G., & Moriarty, T. A. (2010). Introduction: Forging connections among undergraduate writing majors. In G. Giberson & T. Moriarty (Eds.), *What we are becoming: Developments in undergraduate writing majors* (pp. 1–12). Logan, UT: Utah State University Press.

Greenhow, C., Robelia, B., & Hughes, J. E. (2009). Research on learning and teaching with Web 2.0: Bridging conversations. *Educational Researcher, 38*(4), 280–283.

Guttenplan, D. D. (2013, May 12). In Britain, a return to the idea of the Liberal Arts. *New York Times.* Retrieved from http://www.nytimes.com/2013/05/13/world/europe/in-britain-a-return-to-the-idea-of-the-liberal-arts.html?pagewanted=all&_r=0

Kiley, K. (2013, May 7). Price of a bad economy. *Inside Higher Education.* Retrieved from http://www.insidehighered.com/news/2013/05/07/nacubo-survey-reports-sixth-consecutive-year-discount-rate-incre#sthash.YXSlZdy5.dpbs

Lee, C. K.-M. (2007). Affordances and text-making practices in online instant messaging. *Written Communication, 24*(3), 223–249.

Looney, A., & Greenstone, M. (2012, August). A record decline in government jobs: Implications for the economy and America's workforce. *The Hamilton Project.* Retrieved from http://www.hamiltonproject.org/papers/a_record_decline_in_government_jobs_implications_for_todays_economy_an/

Meese, G. P. E., & Wahlstrom, B. J. (1988). Designing graduate programs to prepare the communication leaders of 2000+. *Journal of Business and Technical Communication, 2*, 21–35.

Miller, T. P., & Jackson, B. (2007). What are English majors for? *College Composition and Communication, 58*(4), 682–708.

Reid, A. (2010). The write brain: Professional writing in the post-knowledge economy. In D. Franke, A. Reid, & A. DiRenzo (Eds.), *Design discourse: Composing and revising programs in professional and technical writing* (pp. 254–274). Fort Collins, CO: WAC Clearinghouse/Parlor.

Saller, C. F. (2013, March 4). Wannabe editors: Can you pass a proofreading test? *The Subversive Copy Editor Blog*. Retrieved from http://www.subversivecopyeditor.com/blog/2013/03/wannabe-editors-can-you-pass-a-proofreading-test.html

Sandmann, L. R., Foster-Fishman, P. G., Lloyd, J., Rauhe, W., & Rosaen, C. (2000). Managing critical tensions: How to strengthen the scholarship component of outreach. *Change, 32*(1), 44–52.

Selber, S. (2004). *Multiliteracies for a digital age*. Carbondale, IL: Southern Illinois University Press.

Sharpe, L. T., & Gunther, I. (1994). *Editing fact and fiction: A concise guide to book editing*. Cambridge, UK: Cambridge University Press.

Webster-Wright, A. (2009). Reframing professional development through understanding authentic professional learning. *Review of Educational Research, 79*(2), 702–739.

Weisser, C., & Grobman, L. (2012). Undergraduate writing majors and the rhetoric of professionalism. *Composition Studies, 40*(1), 39–59.

Yeats, D., & Thompson, I. (2010). Mapping technical and professional communication: A summary and survey of academic locations for programs. *Technical Communication Quarterly, 19*(3), 225–261.

Zarefsky, D. (2004). Institutional and social goals for rhetoric. *Rhetoric Society Quarterly, 34*(3), 27–38.

Zerbe, M. J., & DelliCarpini, D. F. (in press). Writing as an art and profession at York College. In G. A. Giberson et al (Eds.), *Undergraduate writing majors: Nineteen program profiles*. Logan, UT: Utah State University Press.

SECTION TWO

New Teaching Models: Adapting Technologies Strategically

CHAPTER 5

Frugal Realities: Hacker Pedagogy and Scrappy Students in an Online Program

Julia Romberger and Rochelle Rodrigo

Old Dominion University's (ODU) Department of English has Professional and/or Technical Writing programs or emphases at all degree levels: bachelor's, master's, and PhD. With this many technical and professional writing programs and emphases, most offered in technologically mediated distance-learning environments, we are implicitly making a promise to provide students, depending on their degree requirements, the opportunity to take various technical or professional writing courses at different levels of complexity or depth (i.e., junior, senior, or graduate courses).

All of these students want to graduate with the academic stamp of *technical/ professional writer* and with the technological experience and knowledge associated with this title. The need to instruct students at various academic levels is further complicated by the background and experiences of the students; some have been working in business and industry for a decade or two, and occasionally they have more experience with a specific application, or type of application, than the instructor. Others are traditional undergraduates, some of whom might come from financially disadvantaged backgrounds and have had little access to technology. Many of our students enter courses with specific goals of learning particular software applications; we cannot meet all of their needs. Even if we had all of the hardware and software available to fulfill students' wants and needs for technological experience and training, we don't have enough faculty with the varied expertise and experiences to fulfill these desires. In short, we are giving up the ability to "credential" our students with a stamp of expertise with a particular set of applications. However, we don't necessarily see that particular practice as sustainable or practicable at most universities with current resource constraints.

This inability to provide students with comprehensive access to hardware, software, and instructor expertise is further complicated by our growing number of distance-learning students. Students in a distance-learning program would need to have the ability to access copies of specific applications themselves. And although ODU does have a Virtual Computer Lab, it does not necessarily have all of the applications a technical and professional writing student might want to learn. Even if all the hardware and software was available to the students, the instructors would also need access and pedagogical expertise to juggle working with the layers of distance learning and communication-mediating technologies during the design and delivery of course instruction along with other professional responsibilities.

While the circumstances of ODU's professional writing program seem less than ideal, we have looked upon them as an opportunity to not only "do more with less" but also to teach students how they might develop metacognitive skills that allow them to transfer abstract thinking about various learning approaches in order to be able to adapt more readily to whatever variety of workplace situation they find themselves in. They are overtly training to understand the whole of the rhetorical situation, including the role that technology plays.

In this chapter, we argue programs that focus on preparing students to understand how to rapidly and collaboratively learn rhetorical usage of technologies produce students better able to adapt to a variety of workplace situations. The pragmatics of austerity in universities and colleges that are increasingly experiencing lean financial times can encourage a hacker mentality, and this mentality can produce scrappy students who in turn understand how to use networks of learning and how to identify potential means of learning new skills and developing new interpersonal networks within their work situation. In the following sections, we will review how the discipline of technical communication has approached student preparation for the profession in the past, offer our definition of and rationale for hacker pedagogy, and describe three sample assignments that exemplify this pedagogy.

SETTING THE STAGE: THE DISCIPLINARY CONTEXT

There is a long-standing debate between academics and practitioners about what is the most valuable preparation for a technical communication student. Academics such as Davis (2001) and Miller (2004) have argued that focusing on technology and a skills-based curriculum can distract from professionalization and a more robust and adaptable education. In 1996, Johnson-Eilola argued that education for technical communicators should focus on preparing them to do symbolic-analytic work—the work of abstracting and manipulating information (p. 253). On the other hand, practitioners have bemoaned the poor preparation of new college graduates, both undergraduate and graduate, for the technologies they will be using in the workplace. In part, this concern can be seen as part of practitioner anxiety about staying ahead of leading-edge technologies to retain their professional relevance (Davis, 2001). This focus can be seen in journals prepared for the practitioner audience, such as *Intercom*. A brief survey of articles from 2011 and 2012 issues shows that on average the journal includes at least two articles explicitly either reviewing or discussing best-practice usage of particular technologies. The variation between placing importance on being conversant with specific applications versus having a strong rhetorical education is just one part of the divide between industry and academia; practitioners in industry are usually very focused, of necessity, on the particulars of their context, whereas academia is attempting to provide a more abstract education meant to be foundational and transferable for students. Specifically, while practitioners want newly graduated peers to start jobs already knowing how to use specific technologies, it is nearly impossible for most academic programs to cover all technologies due to time and money constraints.

Historically, the job ads the market provides often focus on a list of technology knowledge preferred over less tangible competencies like rhetorical awareness; however, there are signs of a shifting trends. Although many job ads still list specific technologies as desired or required proficiencies, especially MS Word (as a brief survey of "technical writer" job ads on ProEdit and USAjobs.gov in May 2013 demonstrates), an increasing number relegate the listing of a specific make or model of a technology to a parenthetical (e.g., listing the type of software and then providing names of applications as examples) or deleting the specific name of an application altogether and just listing categories of technologies like document design, image editing, and learning management systems or content management systems. The shift from listing many specific applications to focusing on the ability to learn the needed technologies can be seen in the following excerpt from a recent job ad: "The ideal candidate will be able to learn new software quickly" (ProEdit, 2013). Job ad language of this type suggests an increasing number of employers recognize it is impossible to know all the available applications that may be chosen to do specific tasks; they want applicants with more flexible technological expertise and experiences. The language possibly demonstrates one of two realities. It may show understanding of the wide variety of application types available to do various tasks. It may also demonstrate that companies in the current economy may need to be more frugal with technology purchasing and thus reenvision their expectations of new hires.

Technical communication programs have traditionally struggled with the ability to offer both the rhetorical training students need and the technological knowledge expected in the corporate world. The technology can be expensive, and academics who already have many responsibilities both inside and outside the classroom often struggle to keep up with the latest technologies. Rather than seeing fiscal and pedagogical constraints as impediments, we claim such constraints can provide opportunity to rethink programmatic approaches to pedagogy. It is an opportunity to enhance technical communication students' abilities to see themselves as symbolic analytic workers (Johnson-Eilola, 1996). These types of workers, which Johnson-Eilola argues technical communicators are perfectly positioned to be, work primarily with content mediated by information technologies of various types.

"Hacker Pedagogy," which we introduce here, works to help students develop an understanding of the process of learning and producing as networked processes that incorporate all of Johnson-Eilola's categories (see also Dicks, 2010). Hacker Pedagogy is a way of developing strong reflective practices that can be adapted to various contexts. In a Hacker Pedagogy, students are encouraged to experiment with technologies and processes; they are expected to collaborate both on learning processes and production; they are explicitly taught to abstract from one technology to the next as well as to consider their assemblage process; and they are encouraged to think of their learning as positioned within a larger, complex system. It also requires recursive reflective practices on the part of the instructors regarding choices about outcomes versus skills and pedagogical goals in light of resources. These choices are very similar to what we ask of the students: How do we meet the rhetorical goals of what we understand to be best educational practices with the contextual constraints that we have?

PRODUCING SCRAPPY STUDENTS: THE PEDAGOGY

The OED defines scrappy as "inclined to scrap or fight; aggressive, pugnacious, quarrelsome" (Oxford Dictionaries, n.d.). *Inclined* implies predisposition to or premeditation of the negative behavior of fighting. However, two of the OED examples of scrappy included more positive connotations of the word: one about fighting for self-preservation and the other about fighting to protect others. Within the context of this program and these assignments, we are asking students to be scrappy by fighting to learn the most within their individual context and design and deliver the best projects within their individual constraints. We want to prepare our students for a number of eventualities, including nonprofit or government jobs wherein their employers will not have the top-of-the-line hardware or software; we want our students to fight through this constraint for their own professional self-preservation and the protection of their projects (and employers or clients). Our scrappy students will recognize that every rhetorical situation will present them with constraints, many of them in terms of access to resources like hardware and software, but that doesn't mean they just quit trying;

instead, we want to graduate students who are *inclined* to see that constraint as a challenge to negotiate, or hack, their way through.

Just as in most "real-world" situations in which technical and professional writers do not have access to a preferred choice, or version, of an application, we don't have all of the hardware or software we would like students to learn, or at least have experience with, in the program. Our students may graduate and work for an organization that either does not own the Adobe Suite or is currently working with an edition that is two versions old. Some graduates may work as independent contractors and only be able to afford to slowly purchase hardware and/or preferred software applications; for instance, instead of using Adobe Photoshop they might need to spend a few years working with the Open Source image editing application GIMP.

In alignment with recognized pedagogical and workplace practices in the field (e.g., Brumberger & Northcut, 2010; Spinuzzi, 2003), our pedagogy emphasizes teaching students to identify, articulate, and then adapt to the given rhetorical situation, which includes understanding what technologies the student has access to while completing a project. Obviously this is the theoretical framework we use to teach technical and professional "writing"; however, we are also asking students to make explicit their own subject position as technology users within the rhetorical situation of both the given assignment as well as the class as a whole. In other words, the learning of the technology is also a rhetorical situation. Therefore, instead of promoting a contrasting pedagogy of skill-and-drill to teach a given technology (which implies everyone is learning the same aspects of the same technologies at the same time—a completely arhetorical method), we ask that students understand both their use and learning of a given technology within the context of the course and their own experiences.

To provide students with the flexibility to insert their own technological agency into a project, the learning objectives must shift slightly. Instead of saying "in this project you will learn how to use Photoshop," project goals in our program generally are more focused on a piece of the rhetorical problem, like "in this project you will learn how to edit and manipulate images." Instead of starting the project by standing up in front of the class and showing them how to manipulate the images (some may already have lots of experience while others have none), we ask students to carefully articulate their experiences, goals, and contextual factors as related to the assignment. In this example of editing and manipulating images, we ask questions like

- What experiences have you had with editing and manipulating images?
- What specific skills would you like to learn about editing or manipulating images?
- What image editing software do you have access to?
- What support resources do you have access to?

while introducing the assignment and explaining the philosophy to the students regarding how they will identify their own learning goals and access to software and learning support materials. For example, a student who had a decade of professional experience with the basic Adobe Suite asked, "So, I've wanted to start becoming more comfortable using my smartphone to take and edit pictures. You are saying I can set that as my goal for this assignment?" For someone who already had a lot of experience with Adobe Photoshop, learning to take and edit images using different application on a smartphone was a perfect goal!

Within this philosophy, it becomes the responsibility of instructors to construct environments and assignments that are flexible enough to allow students to flesh out the rhetorical details of both the assignment and their learning. Instructors also need to help them reflect on their own prior experiences, identify and articulate their desired learning goals, as well as find and locate access to software and software learning support. Lastly, instructor responsibilities include encouraging students to push themselves beyond the boundaries of what they can comfortably access and learn. By doing this, students learn to work through and beyond cognitive dissonance to create productive knowledge about rhetorical and pragmatic approaches to technology. In short, the program asks that students put themselves in complex situations in which they have to explicitly, even *aggressively*, learn something new to achieve the objective.

Although many people consider hackers bad people, and most hackers *do* break the law, Turgeman-Goldschmidt's (2008) ethnographic study found that "both good and bad hackers explain their practices in terms of: 'breaking boundaries,' 'shattering conventions' and 'doing the impossible'" (p. 391). One of the participants in the study suggested

> a positive connotation of the term hacker: a computer technology expert who "does the impossible," proves his/her ability and superior expertise, and belongs to an elite subculture of experts in the field who are leading society toward a better technological future. (p. 387)

In an article comparing contemporary hackers to trickster gods, Nikitina (2012) claims that hackers focused on "more than developing new methodologies"; instead "the hackers' mind is engaged in the process of rearranging the givens of existing systems" (p. 143). We draw parallels between ourselves, our students, and hackers—individuals working within a given scenario and creatively using what they have access to achieve the goals.

In *A Hacker Manifesto*, Wark (2004), like many writing instructors across the United States, emphasizes *process*:

> Hacking is the production of production. The hack produces a production of a new kind, which has as its result a singular and unique product, and a singular and unique producer. Every hacker is at one and the same time producer and product of the hack, and emerges as a singularity that is the memory of the hack as process. (p. 158)

Hacking as pedagogy produces itself. Students as hackers hack their own learning. Hacking is both the pedagogy and the learning. And if we want the learning to transfer to future scenarios, not only do we, as hacker instructors, have to be critically reflective and transparent, we need for students to become meta-aware of their own hacking as learning through critical reflection as well. Students who take up a hacker agency become scrappy in their critical adaptive ability to assess the resources in a particular situation in order to accomplish a rhetorical task.

Just as students must be aware of the learning objectives and their access to and experience with the technology prior to an assignment, students also need to reflect upon how they completed the assignment, what they learned, and how they learned it. Each resource a student accesses to help design and develop the course project leaves a *trace* of what the student learned and how s/he learned it. Within the process of completing a project, students will definitely access the technologies they are using; however, they might also go to the instructor, other students, technology support websites and videos, friends and family, ad infinitum.

THEORIZING HACKERS AND HACKER ASSIGNMENTS

We turn to Latour's (2005) Actor-Network-Theory (ANT) to help articulate various actors, or actants (p. 54), and the connections (or *traces*) between them, within a Hacker Pedagogy. ANT is a research methodology developed out of sociology, which defines "social" as a "type of connection between things" (Latour, 2005, p. 5). ANT acknowledges that both people and things (including technologies) can be participants/actants in social connections; specifically, ANT sees most actants as active mediators (when they interact with something, they change it) (p. 39). ANT sees the social only through social connections, specifically the traces of action between the connections. These types of participants and connections between them help define Hacker Pedagogy. All actants in Hacker Pedagogy—students, faculty, assignments, technologies, support structures, and such—are active mediators, constantly moving in a fluid state of connections with different ontologies. This is why faculty and students must be both critically reflective and transparent in Hacker Pedagogy; they must be both aware of as well as *trace* their own status as active mediators and how their actions change other actants.

Each student within the hacker pedagogical situation is an actant who brings his/her own already developed learning, writing, and technology support networks to the writing situation. Instead of creating a circumscribed assignment with highly specific parameters with only one approach for the student to arrive at the learning outcomes that assumes similar students with similar experiences and resources (i.e., assignment as *intermediary*) (Latour, 2005, p. 39), in a Hacker

Pedagogy, faculty (as active mediators negotiating their own history and context) create assignments that allow students to pick and choose among self-identified (or co-identified with the instructor) actants of technology and/or deliverables (i.e., assignment as *mediator*) (Latour, 2005, p. 39).

In this way, each student is allowed to develop their own fluid, active learning processes, with the endpoints being new resources discovered during processes of learning. These resources become new actants in students' ongoing development of learning, writing, and technology support networks. Like Latour (2005) in defining how to interact with actants in ANT, we acknowledge that the actants best know their own network in all its fluidity and complexity; we want them to be able to articulate the value of their learning network and how it came into being so that they are able to identify and use what actions worked best for them and what work of assemblage was beneficial. This mindfulness about their assemblage processes is a transferable concept that they can then take on the road with them to better allow them to read and adapt to a new context. Hacker Pedagogy adopts a mindful, reflective, participant-as-researcher, self-aware ANT. Just as Latour states that ANT scholars must write, must "make reports" (p. 122), students in a Hacker Pedagogy must reflect, write down, and *trace* their learning and production processes, and these reflections must be able "to register differences, to absorb multiplicity, and to be remade for each new case at hand" (p. 121).

So to be specific about what a Hacker Pedagogy is, we see such a pedagogy as occurring at the point of exigence where the need for a new communication is identified. At this point, an instructor will assist students in learning to implement critical reflective practices that will lead to adaptive capacities. The underlying goal is to encourage students to understand that these are transferrable practices. This work of critical reflection is done through consideration of

1. who the actants are (including the technologies and material constraints);
2. what the rhetorical goals are; and
3. where the resources are for learning, hacking, and completing the communication task.

This occurs in a reciprocal process that continually refines this knowledge through reflection during the process of production until the communication is released into the larger discursive situation to meet its purpose.

The things that distinguish Hacker Pedagogy are the focus on continual critical reflection and adaptation as well as the focus on technology as an actant that inform rhetorical choices. We are overtly encouraging them to operate in the disruptive way that Spinuzzi (2002) has identified.

Asking students to take responsibility for their own learning is not new as a pedagogical strategy. In fact, it aligns with many social constructivist pedagogies that emphasize student learning versus instructor teaching (Barr & Tagg, 1995), giving students power to decide what they are learning (Weimer, 2002), collaborative learning and distributed knowledge making (Kalantzis & Cope, 2008), as well as learning how to learn (Fink, 2003). A Hacker Pedagogy promotes lifelong learning by having students identify what they want and need to learn, why they need to learn it, and providing them with the confidence to find resources to help them learn it. And as any professional writer or technical communicator will tell you, committing to the profession is a commitment to lifelong learning of the content production, management, and publication technologies used. Scrappy professional and technical writing graduates who hack their way through the following types of projects will be well prepared for a market that is continually changing.

HACKING ASSIGNMENTS: THE EXAMPLES

This Hacker Pedagogy does not work without both students and faculty being critically reflective and transparent. We, the faculty, must first be critically reflective of our pedagogical objectives within the framework of our austere context. Just as our Hacker Pedagogy asks students to carefully articulate their learning goals and context, performing Hacker Pedagogy requires that we are transparent in our practices. We want our students to recognize the program's austere situation, and we want them to recognize their current and future selves within constrained circumstances. By being transparent with the students, our Hacker Pedagogy models the Hacker Philosophy we want them to learn within the program. We are asking the students to replicate our reflective practices that look at the authors/agents/actors within a rhetorical situation, how the pathways toward achieving the goals and purposes of these actants are informed by constraints and affordances, and what the resources to be used to adapt to the rhetorical situation might be. As instructors, we are always considering carefully

- who our student population is generally and specifically (and allowing them to tell us within their assignments),
- what might be the constraints and affordances of the context at several levels (that of the university, the distance-learning infrastructure, the program, and the classroom),
- what are the general learning outcomes (programmatic, course specific, assignment), and
- how our assignments can be adaptive to specific students and their constraints.

All of these are precursors to overt conversations about transference of adaptive practices as an outcome of reflection as a major rhetorical goal.

Below, we share a few of the assignments at the undergraduate and graduate level that we feel exemplify the Hacker Pedagogy approach by demonstrating the characteristics listed above. They each include space for critical reflective practices and encourage adaptive behaviors by helping students carefully consider all actants, rhetorical goals, and resources. While we give the level for which the assignments are articulated, each of these assignments can be scaled to work at a different level. These assignments are also rendered in such as way as to make them conducive to the realities of a program that has just as many of its courses at a distance as face-to-face. The assignments are not highly dependent upon access to *A* particular technology in *A* particular context. They also can work well in a sequence with the first assignment setting up students to participate in Hacker Pedagogy by asking them to take responsibility for their learning and share what they learn. The second assignment focuses on helping students understand technology as an actant. The final assignment is a core example of a hacker multimedia project.

As part of Hacker Pedagogy, all of the assignments require a significant reflective component. All of the assignments ask students to reflect both at the beginning and end of the assignment. Early reflections ask that prior to the assignment students assess their rhetorical situation, especially their position as author/actor/agent as both a learner in the class as well as an author within the project. This will require students to acknowledge layering of rhetorical components (such as an intended audience as well as instructor as audience) and identify their individual experiences as related to the assignment goals and their learning goals associated with the particular assignment. As part of their self-analysis, students will need to identify their resources, both technological and knowledge based.

After completing the assignment, students will need to complete a similar reflection that asks them to account for their process and product. Students need to reflect upon their approach to the final product—what they did and why they did it. Students also discuss their process—their learning as well as their negotiation of their resources while completing the project. And since contexts change, students should also re-reflect upon the rhetorical situation to see if elements shifted and how that impacted their product, process, and overall learning. In short, to be an assignment that fits within a Hacker Pedagogy, the assignment must construct a scenario that invites students to identify, articulate, and adapt to a rhetorical situation for both the assignment project as well as the technology learning. The assignment must also ask students to trace their learning by critically reflecting upon their individual experiences before, during, and after the assignment.

Assignment #1: Understanding "Expertise"

In short, the goal of this assignment is to have students share the responsibility of being experts on or about different technologies. Each is either assigned and/or picks a technology that is relevant to the course, does research on it, and then reports the results to the remainder of the class. We use different versions of this assignment in multiple classes. In an upper-division Digital Writing class, it functioned to help give students different options for learning to edit audio, video, and images. In a graduate course, it functioned to report on different programming languages that students would then pick from to spend some time learning.

Assignment Goals

After completing the assignment, the goal is for the student (or a group of students) to take ownership over a particular technology and then feel more comfortable using the individual technology as well as more confident learning new technologies. While doing the research for the assignment, students should learn how to identify the affordances and constraints generally associated with a given technology. They will also develop a network of resources to help learn about and troubleshoot new technologies they may encounter.

Assignment Prompt

Folks producing digital new media need to be at least explicitly aware of, and/or extremely familiar with, these various technologies. Fifteen weeks is barely enough time to learn about one of these technologies and play around with it; however, five weeks is enough time to research one of these technologies. One of the best ways to learn about a new technology is to teach it to someone else.

As a class, you will share the burden of learning about and teaching these various technologies. You will construct reports, both written and oral, to teach your classmates about these technologies. For your assigned technology, you will need to research and produce (a) wiki page(s) (and an oral report) about the following:

- What is it? What does it do/enable? (Short enough to be easily consumed, detailed enough to be useful.)
- History of it. Where/how/why was it developed? (Short enough to be easily consumed, detailed enough to be useful.)
- Annotated list of hardware and/or software requirements (and versions of software).
- Annotated example list/links (at least 3) of websites/projects using this technology.
- Annotated resource list/links (at least 4) for learning and developing/designing using this technology. Consider people, places, and other things (especially material on the Web) as possible resources. I do not expect you to provide a detailed critique of these resources as if you've used them, I'm just looking for a brief summary of what the resource includes and if you've found some reviews; include summaries of the reviews as well.

Since this assignment will be submitted in a wiki page, be sure to take advantage of the affordances of the technology. You will want to be sure to include relevant hyperlinks and representative images; you might also consider embedding informational or instructional videos as well.

Scaffolding Activities

- Proposal Memo: In memo format, tell the instructor what technology you will be researching and why you are interested in researching it. Provide a work schedule for how you will complete the assignment with ideas on where you will find the information and resources you need. [At minimum, this assignment allows instructors to know who is doing what technology; instructors might take the time and energy to provide detailed feedback to help students with their work. Instructors might also have students share ideas of when and where they will be doing their work.]
- Resource List: Submit a list of resources that you have found so far. Make sure you have found at least two of each listed below:
  –General information resources about your technology
  –Examples of texts produced with the technology
  –Educational resources
  [Students generally need help finding examples and authoritative educational resources; starting them early gives them time to get help from one another, the instructor, and others.]
- Complete Draft: Have a full draft of your wiki page designed and developed.
- Peer Review: Read and respond to at least two of your (assigned) classmate's wiki pages using the following three questions and the grading rubric to inform your response:
  –What did you like? Why? How might your classmate repeat what they did well in future projects?
  –Where were you confused? Why? Ask questions.
  –Where would you like to know more? Ask questions.

Assignment #2: Analyzing Interfaces

The driving goal behind this 3–4 week assignment is to enable students to read interfaces as rhetorical spaces. We have taught this assignment at both the undergraduate and graduate level, changing the readings and the materials available with it. This report encourages them to think about the process of composing at a more complex level, which includes its materiality—the actual act of working with a program to create a visual or verbal text—and to think about the "interface" of their own documents of paper, PowerPoint, or Web page, along with the delivery and interactive structures. This version is for an undergraduate Introduction to Digital Writing class, wherein the unstated goal for many of our students is to be comfortable with technologies that they are not regular users of. We begin with an explanation of interface design and the concept of discourse communities and how they appear in the interface in order to show complexity and multiplicity of discourse expectations. This allows students to understand how such interfaces are complicated communication spaces with overlapping discourse expectations.

Assignment Goals

The goal of this assignment is for students to be able to "read" an interface and apply previous knowledge and expertise with other interfaces to new applications in a conscious fashion. Students will better understand that the choices designers made regarding these interfaces were communication choices attempting to reach particular audiences.

Assignment Prompt

The function of this project is to encourage you to think about interactivity, the metaphors used by applications, and the audience for which these applications are created. All of these things shape the way that users engage with the writing technology and utilize them for composing tasks. Feel free to choose an application you want to learn more about or an application that you have used in the past that puzzled you or for which you feel you are the primary audience. You can look at a photo editing application, a bulletin board, an email program etc., but it should be discrete. For example, if you are looking a social networking site, you should look at one part of the site like the chat room. For this project, you will create a very brief proposal that must be approved by the instructor before you begin your short report. Additionally, you will write a short report addressed to the instructor that uses the readings to discuss the metaphors being used by the program to facilitate those tasks and discuss what discourse community would be familiar with those metaphors.

Scaffolding Activities

Since students have not traditionally been taught to read interfaces, there are various points of entry to this assignment that the students will understand and value.

Class Discussion and Real-World Analysis

One way to begin framing the value of interface critique is to show the practical application of such inquiry to students. An examination of the history of Facebook's redesign and why the users should have been able to vote their acceptance of it is a useful talking point. Many users had been unhappy with changes in layout and the privacy settings, and how privacy settings must be accessed in order to reset them. This history allows the students to think about the interface of Facebook from within their personal discourse community as Facebook users, which many of them are. They can better understand how an interface impacts those who use it (for some recent examples, see Kincaid, 2010; McCullagh, 2010; Peters, 2009; Swartz, 2010). Students can then be shown how interface critique can assist them in learning applications that are unfamiliar to them by understanding how categories in the critique are also heuristics. For example, we discuss the assumptions that DreamWeaver has about its user as experienced Web designers, which stands in contrast to most students' experiences as first-time users. They do not have access to a lot of the particulars of that discourse community, but by identifying it through understanding the primary purpose(s) and the metaphors that the interface draws from that discourse community, they can learn to read the interface. They are then asked to reflect upon their own regular technology usage and how those technologies fit them as users.

Introduction to Terminologies

These help students articulate what they see during their rhetorical analysis. We discuss user, designer, programmer, toolbar, menu, palette, ribbon, icon types, and visual metaphor. After an introduction to these terms, we discuss slippage between types and the problems with using real-world metaphors as a cognitive map for the foundation of interface design drawing upon Laurel's (1991) work as well as students' own digital experiences. In order to assist students in understanding how discourse communities influence the design of interfaces, the assignment builds off the terminology of the lecture materials to discuss layout and functionality basic to nearly all software. It helps to introduce students to the preference settings in common applications and teach them that nearly all applications and operating systems have such settings. Then we demonstrate to students how to change what is placed in their standard toolbars or ribbons and that they can download icons for files, folders, and programs so that they can change some of the metaphors to suit their own subjectivities. Showing students how an application like Adobe's CS5 and CS6 versions of Flash contain multiple layouts targeted toward specific user purposes helps introduce them to the concept of discourse community.

Discourse Community Definition Building

As we discuss interface, metaphor, and icon type, students work on their own definitions of discourse communities based upon reading a summary of Swales' Genre Analysis (1990, ch. 2) and their own research into online discourse communities they have interest in, which is a one- or two-class activity. The focus is on specific parts of Swales' definition of discourse community—agreed-upon goals, genres, and specialized lexicons that discourse communities use and the reasons behind this usage. A useful and fun example that really makes the point for students is the icanhascheezeburger.com site, which features lolspeak, a language developed by site members for reasons specific to their primary goal—humor.

Analysis Worksheets

The final in-class assignment is a pair of worksheets. First, students complete worksheets in which they use the terminology they have learned to discern what iconography and the default workspace window size says about how the presumed audience is expected to use the application. This worksheet is a team-developed rhetorical analysis that allows students to put into use what they have learned and get feedback from me on their understanding of the terms.

The second worksheet, provided after the topic/application proposals have been approved, is a data-collection tool for the students to use individually to compile information for their report. Space is provided for peer feedback in addition to instructor feedback. These two worksheets lead up to an assigned Interface Rhetorical Analysis report.

Assignment #3: Producing Multimedia

This assignment resembles many of the multimodal and/or multimedia production assignments found in classes ranging from First Year Composition to graduate-level digital writing courses. The goal of the assignment is to give students specific practice and advancing skillsets in working with a specific media (in the case below, video). Hacker Pedagogy requires constructing a project scenario that allows for the work to be completed in a variety of ways

Assignment Goals

After completing the assignment, students will be able to design a short video that accounts for coordination between at least one audio and one video track. Students will be able to develop a video using video editing software that allows separate editing of audio and visual tracks. [This particular assignment was given to a joint upper-level undergraduate level course; it would be easy to increase or decrease the level and complexity of the assignment based upon the assignment's learning goals, especially by adding multiple sources, requiring various audio and/or visual tracks, and/or requiring specific rhetorical devices like certain style editing.]

Assignment Prompt

You will construct a video advertising a book (e.g., http://youtu.be/x0Enh Xn5boM). You will do all your video editing in one of the following applications: iMovie, Windows MovieMaker, or propose your own. I hope you will upload your final project somewhere like YouTube and/or Vimeo. If you do not publish it to the Web, you will need to get the final project to me somehow (NOT via email).

To help with the assessment of the project, and to demonstrate your learning, you will also submit a cover letter that includes

- the full bibliographic information about the book you are advertising along with a discussion of a specific audience you are trying to "sell" the book to;
- (if you publish it) an embedded video or link to the final photo essay set;
- discussion of where and how you found the original images, videos, and/or music;
- discussion of the copyright implications of the images, videos, and/or music you used and how you followed copyright and/or included citations; and
- detailed discussion of the process used at each step:
  —pre-production,
  —production, and
  —post-production; as well as
- discussion of what processes you took to learn the video editing application, what you learned, and what more you hope to learn in the future.

Scaffolding Activities

Pre-production:

- Select Book: finalize book selection along with detailed discussion of the specific audience; brainstorm tone and style ideas; brainstorm image ideas.
- Select Technology: finalize video editing software selection; discuss your previous experience with any/all video editing software and how/why you are interested in learning about the one you selected as well as how/where you are going to access the software during this project.
- Learn the Technology: read and/or watch a tutorial on the video editing software you selected; play with the software a bit; reflect on what you learned and how you learned it.
- Design Video: detailed storyboard of your video with discussion of audio as well as where/how you will get the original content.
- Schedule Process: production schedule (what/where/how you will get find, shoot, and produce everything for the project). Include date, what you are doing, where you are doing it, requirements (materials, permission, etc.).

Production:

- Update #1: update of what you have done, where you are, what you need to do, how it is going, problems, concerns, etc.
- Update #2: update of what you have done, where you are, what you need to do, how it is going, problems, concerns, etc.

Post-production:

- Complete Draft: Have a complete draft of your video ready to share with your classmates and get feedback.
- Peer Review: Watch and respond to at least two of your (assigned) classmate's videos. Using the following three questions and the grading rubric to inform your response:
  —What did you like? Why? How might your classmate repeat it?
  —Where were you confused? Why? Ask questions.
  —Where would you like to know more? Ask questions.

CONCLUDING REMARKS

The above assignments demonstrate Hacker Pedagogy in that their structure takes into account contextual and student differences in affordances and constraints. Each assignment models transparent understanding of necessary adaptive practices and encourages understanding that such adaptive practices can be transferable. They all reinforce critical reflection as process and product of student learning and a space in which students and faculty can think about how adaptive practices are generated.

Hacker Pedagogy benefits both faculty and students. For faculty, it allows them to work within the increasingly austere institutional contexts in which they find themselves and still support professional and technical writing students in a pedagogically sound and ethical manner. Hacker Pedagogy benefits students by helping them conceptualize the building of a network of different technologies and support structures to achieve a variety of workplace goals. Because these students will understand all the moving parts of various rhetorical situations, which include technologies, resource adaptation, and usage, they will know how to best work within the constraints given to them. They will be able to continually construct fluid networks of resources and agents in order to accomplish their professional goals.

REFERENCES

Barr, R. B., & Tagg, J. (1995). From teaching to learning: A new paradigm for undergraduate education. *Change, 27*(6), 13–25.

Brumberger, E. R., & Northcut, K. M. (2010). Resisting the lure of technology-driven design: Pedagogical approaches to visual communication. *Journal of Technical Writing and Communication, 40*(4), 459–471.

Davis, M. (2001). Shaping the future of our profession. *Technical Communication, 48*(2), 139–144.

Dicks, R. S. (2010). The effects of digital literacy on the nature of technical communication work. In R. Spilka (Ed.), *Digital literacy for technical communication: 21st century theory and practice* (pp. 51–81). New York, NY: Routledge.

Fink, L. D. (2003). *Creating significant learning experiences: An integrated approach to designing college courses*. San Francisco, CA: Jossey-Bass.

Johnson-Eilola, J. (1996). Relocating the value of work: Technical communication in a post-industrial age. *Technical Communication Quarterly, 5*(3), 245–270.

Kalantzis, M., & Cope, B. (2008). *New learning: Elements of a science of education.* Cambridge, UK: Cambridge University Press.

Kincaid, J. (2010, February 4). Facebook begins rolling out latest redesign (pics). *Tech Crunch.* Retrieved from http://techcrunch.com/2010/02/04/facebook-begins-rolling-out-new-redesign-pics/

Latour, B. (2005). *Reassembling the social: An introduction to actor-network-theory.* Oxford, UK: Oxford University Press.

Laurel, B. (1991). *Computers as theatre.* Reading, MA: Addison-Wesley.

McCullagh, D. (2010, June 16). Advocacy groups: Facebook privacy changes not enough. *CNET News*. Retrieved from http://www.cnet.com/news/advocacy-groups-facebook-privacy-changes-not-enough/

Miller, C. (2004). A humanistic rationale for technical writing. In J. Johnson-Eilola & S. A. Selber (Eds.), *Central works in technical communication* (pp. 47–54). New York, NY: Oxford University Press.

Nikitina, S. (2012). Hackers as tricksters of the digital age: Creativity in hacker culture. *The Journal of Popular Culture, 45*(1), 133–152.

Oxford Dictionaries. (n.d.). *scrappy* (2). Retrieved from http://www.oxforddictionaries.com/us/definition/american_english/scrappy

Peters, M. (2009, March 27). Facebook's new format takes away from its purpose. *The Huffington Post.* Retrieved from http://www.huffingtonpost.com/meghan-peters/facebooks-new-format-take_b_179803.html

ProEdit. (2013). 90-day apprenticeship: Assistant information developer [Job advertisement]. *ProEdit: words+people*. Retrieved from http://proedit3.rssing.com/chan-15938540/all_p1.html

Spinuzzi, C. (2002). Toward integrating our research scope: A sociocultural field methodology. *Journal of Business and Technical Communication, 16*(1), 3–32.

Spinuzzi, C. (2003). *Tracing genres through organizations: A sociocultural approach to information design*. Cambridge, MA: MIT Press.

Swales, J. (1990). *Genre analysis: English in academic and research settings*. New York, NY: Cambridge University Press.

Swartz, J. (2010, May). Early reaction positive to Facebook privacy changes. *USA Today.* Retrieved from http://content.usatoday.com/communities/technologylive/post/2010/05/facebook-acted-quickly-decisively-in-addressing-latest-privacy-f/1#.U2RVNPldXh4

Turgeman-Goldschmidt, O. (2008). Meanings that hackers assign to their being a hacker. *International Journal of Cyber Criminology, 2*(2), 382–396.

Wark, M. (2004). *A hacker manifesto*. Cambridge, MA: Harvard University Press.

Weimer, M. (2002). *Learner-centered teaching: Five key changes to practice.* San Francisco, CA: Jossey-Bass.

CHAPTER 6

Service-eLearning in the Online Technical Communication Classroom: Keeping Our Pedagogies Relevant in an Age of Austerity

Tiffany Bourelle

> The future's uncertain, and the end is always near.
> — Jim Morrison (1970)

While the end may not be as near as Jim Morrison claimed in *Roadhouse Blues*, in terms of the future, uncertainty may be putting it mildly. As the United States continues to climb slowly out of the recession, the job market for new graduates remains bleak. In 2009, some 1.5 million U.S. undergraduates received bachelor's degrees, while an estimated 1.9 million U.S. workers who held bachelor's degrees or higher remained unemployed (Eisner, 2010, p. 27). While this number may seem concerning, those with an education are faring better in the job market than those without. Eisner (2010) asserts that, typically, individuals with bachelor's degrees or higher tend to have "stronger labor force participation rates than any other educational group" (p. 28). These trends keep enrollment high, with the National Center for Education Statistics reporting 8 million students attending college in 2010. Unfortunately, time spent in the classroom can sometimes be seen as time not spent working and therefore, money lost. Enter online education: an economic solution for universities experiencing budget cuts and students wanting to go back to school to raise their current status. In fact, approximately two-thirds of universities across the country have increased their online courses to accommodate more students while functioning under tighter budgetary constraints (Parsad & Lewis, 2008, p. 2).

Specific to the field of technical communication, entire programs are shifting online, with many offering full degrees from a distance. However, administrators may think of online education as a quick fix to budgetary concerns without considering the ramifications the online environment can have on student learning and even a teacher's pedagogy. Quite simply, the use of technology can have a tremendous impact, not only influencing the design of the course, but also affecting the student experience. For instance, research suggests that the use of digital tools such as blogs can aid in the unification of the class community, strengthening and enriching the learning environment (Yang, 2011, p. 123). From the course design to course tools used, even down to what types of projects students create, technology can be the most influential factor in online education and must not be overlooked as we continue to move full-force into the online world.

Technology impacts our pedagogies within technical communication in other ways as well, with students in these courses creating multimodal projects like wikis, documents that Carliner (2009) suggests are similar to the digital texts technical writers will be asked to produce in the workplace (p. 28). Other scholars agree, with a 2012 special issue of *Technical Communication Quarterly* focusing solely on the creation and assessment of multimodal documents (Ball, 2012; Manion & Selfe, 2012). These digital texts represent a marked difference in the projects students created merely 25 years ago on a typewriter—an ancient machine by today's standards. Courses with a multimodal focus are becoming more prevalent, encouraging instructors to think holistically, rhetorically, and creatively about communication in digital environments. The wave of the future is upon us, changing even now as this chapter goes to print. How can we provide a lasting education for students within the evolving landscape of academia?

Service-learning, a well-known pedagogy in our field, can prove advantageous in an online classroom, as it allows distance education students to interact with technology in new ways while engaging them in civic activities. In order to meet the demands of online education while encouraging skills necessary for the workforce, implementing service-learning in the online technical communication classroom would be a viable solution. Like traditional service-learning, service-eLearning, as it is commonly referred to, provides an authentic learning experience that teaches students real-world skills, bridging the gap between the classroom and the workplace. However, unlike its counterpart, service-eLearning incorporates technology as the foundation for the pedagogy. Dailey-Hebert, Donnelli-Sallee, and DiPadova-Stocks (2008) describe service-eLearning as "an integrative pedagogy that engages learners through technology in civic inquiry, service, reflection, and action" (p. 1), with technological literacy as a cornerstone for the pedagogy.

In this chapter, I provide an example of service-eLearning implemented in a fully online technical communication course taught at Arizona State University (For more information regarding setting up an online service-eLearning course, see Bourelle, 2014). I argue that service-eLearning in the technical communication classroom can be a way to meet the challenges of technical communication, including the demand for more online classes and increased need for advanced skills in a field in which technology is constantly evolving. Specifically, I discuss the task of providing authentic, real-world experiences for distance education students that are similar to face-to-face (f2f) students' experiences, highlighting assignments and activities used to enhance learning of course objectives in an online course, including structured discussions, collaborative projects, and reflection practices, while meeting the programmatic outcomes as established by Arizona State. As I illustrate, service-eLearning can prepare our students for the digital, electronically linked future, teaching students critical thinking and problem-solving skills that will remain beyond the shelf life of most technological programs. Drawing on the established principles and practices of service-eLearning, I connect the goals of the pedagogy to technical communication, providing a theoretical basis as well as practical advice for instructors striving to keep their curriculum fresh and innovative in the face of institutional and programmatic constraints. Finally, I address the challenges of building a service-eLearning curriculum and sustaining pedagogical practices when resources are scarce at the programmatic level.

INSTITUTIONAL CONTEXT

I first taught a service-eLearning course as a lecturer at Arizona State University, where the technical communication program has a unique and interesting background. When the program was first implemented, most of the courses were taught face-to-face; however, these courses were eventually shifted almost completely to online. The initial reason for the shift was mostly because the online environment could offer greater flexibility for students and instructors (B. Maid, personal communication, May 5, 2013). The technical communication program is housed on the Polytechnic campus, which is a great distance away from the main campus in Tempe, and many students and instructors found it difficult to drive the long distance, especially since many were located in and around the sprawling Phoenix area. When the program was moved online, classes filled more than they did in the f2f format, and the program thrived.

According to Maid and D'Angelo (2013), all was well in the department until 2009, when, because of budget cuts, the "independent writing program was placed in a larger quasi-unit called Interdisciplinary Humanities and Communication" (p. 11), causing the number of majors to grow from 30–35 to well over 100 students. This rapid increase was disconcerting, and administrators and teachers alike were concerned about maintaining a quality education for their students as the program experienced growing pains. To meet the budgetary demands while still providing a quality education, the program moved to a second generation of online delivery, what Maid and D'Angelo call "Online 2G," or courses with set syllabi, outcomes, and assignments, in order to "maintain consistency across all sections" (p. 12). Unlike the courses before, which were mostly f2f courses simply shifted to online, these new courses were designed specifically for online delivery with online learning styles in mind.

As a lecturer, I have taught both types of courses: the shifting of a f2f class to online and the predesigned Online 2G. While there was no specific training for either type of course mentioned, I have a bachelor's in technical communication and worked for more than a decade as a technical writer; therefore, I was first asked to teach an Online 2G course because of my prior experience and because the Online 2G course would let me get my feet wet within the program. I had also taught a Writing in the Professions course on the Tempe campus, which was based on technical communication principles. In this f2f course, students wrote many technical documents and multimodal texts, including fundraiser proposals for local nonprofits; in addition, they actually helped the nonprofits implement the events and raise money. This specific course laid the groundwork for the service-eLearning course I taught in the technical communication program for the Poly campus, Principles of Technical Communication, as this would be based on similar pedagogical principles. This course was not an Online 2G with a structured curriculum set for me, and I was somewhat clueless as to how to shift service-learning into the online realm; therefore, I started with the foundational theory and practices that were commonplace within service-learning and service-eLearning.

SERVICE-LEARNING AND SERVICE-ELEARNING: THEORY AND PRACTICE

Extending from the work of Dewey, service-learning is rooted in the principles of experiential learning wherein students gain real-world experience by working directly with businesses or organizations within their community. In "A Service-Learning Curriculum for Faculty," Bringle and Hatcher (1995) define service-learning as

> a course-based, credit-bearing educational experience in which students (a) participate in an organized service activity that meets identified community needs, and (b) reflect on the service activity in such a way as to gain further understanding of course content, a broader appreciation of the discipline, and an enhanced sense of personal values and civic responsibility. (p. 112)

The pedagogy finds a natural home in technical writing, in which instructors strive to teach practical workplace skills. Since Huckin (1997) first prompted instructors to integrate the pedagogy, service-learning has been adopted and adapted by numerous teacher-scholars in the field (Rehling, 2000; Sapp & Crabtree, 2002; Scott, 2006) in an attempt to bridge the gap between academia and the workforce while simultaneously developing a critical consciousness (Bourelle, 2012, p. 185). More recently, our field has seen a rise in multimodal projects developed for clients or nonprofit agencies, with Walsh (2010) discussing how students' creation of wikis for nonprofits can better prepare students for the rapidly changing field while encouraging them to work within their community (p. 184).

Because the shift to a fully online education is a relatively new trend for many universities, the scholarship surrounding service-eLearning is somewhat limited, especially in the field of technical communication; however, scholars in other disciplines have reported success. For instance, in "Extreme Service-Learning (XE-SL): E-service-Learning in the 100% Online Course," Waldner, McGorry, and Widener (2012) discuss two case studies of courses taught fully online, one in public administration and the other in marketing, claiming that the pedagogy deepened student engagement while compensating for the lack of student interaction that is sometimes present in online courses (p. 849). Hill and Harris (2008) found that the learning management system used within a hybrid professional writing course can affect the service project in numerous ways, suggesting that Blackboard, specifically the features of Group Pages and File Exchange, made the collaboration between students and service partners more productive (p. 83). Other reports of successful practices include service-eLearning used in healthcare professions courses (Kahn et al., 2008) in a language arts class (Strait, 2008), and within business management (Hoover, Casile, & Hanke, 2008), with many instructors using tools such as video- conferencing software and wikis to foster communication among students and service partners.

MEETING PROGRAMMATIC OUTCOMES

When teaching a fully online course in technical communication, I have found and reported success with service-eLearning when the curriculum is carefully planned before class even begins (Bourelle, 2014). For example, in the Principles of Technical Communication online course, I knew I wanted to integrate a service-learning pedagogy; therefore, my first step was to choose a local nonprofit my students could work with throughout the semester. Choosing a partner may seem easy; however, many nonprofits may believe the online element will require a greater time commitment than they are able to provide. When I approached three nonprofits I had worked with in previous semesters, only one agreed to work with my class: the Peace Frame Initiative, an organization dedicated to the eradication of stereotypes regarding under-represented cultures, focusing primarily on Muslims and Latinos. I set up a meeting with the director to discuss their needs and the curriculum for the course. While I had flexibility with assignments, the overall curriculum had to be based on the program's outcomes, which were variations of the Writing Program Administrator Outcomes for First-Year Writing, including Rhetorical

Knowledge, Processes, Critical Thinking, Reading, and Writing, and Knowledge of Conventions (see Appendix for complete list of program outcomes).

To ensure a productive relationship during the service project, I suggest discussing the program outcomes and objectives of the course with the nonprofit, but letting them have a voice in choosing the assignments. Together, the director and I decided what documents the students would create and how they could fit into the programmatic outcomes. We agreed the students would create three major projects, which included a recommendation report, a fundraiser proposal, and a multimodal document; the variations of genres and formats would meet the outcome of Knowledge of Conventions, and writing for a "real" audience would help students meet the outcome of Rhetorical Knowledge. For the recommendation report, students individually researched three fundraisers the nonprofit could easily implement, evaluated them, and chose one that seemed most feasible. Within groups of four, students selected one fundraiser to research and propose to the nonprofit. During this project, each group nominated a team member to Skype with the nonprofit to minimize the director's time commitment, and I met with the students individually when necessary. Each week, the students discussed ideas via a synchronous discussion board, helping each other along the way and participating in peer review; working together and receiving feedback from their peers also helped students learn the outcome of Processes.

In a service-eLearning class, students need to not only learn to work with their peers, but they must also form a relationship with the nonprofit as well. This relationship between the nonprofit and the students must be reciprocal—the nonprofit can certainly gain documents that will help their endeavors, but the students also need to gain knowledge of their community that will last beyond academia. Dubinsky (2002) claims that service-learning's lasting effect depends upon the balance between service and learning; to achieve this balance, students must learn to work *with* the organization—to see themselves as true partners working toward the same goals (p. 64). To foster this partnership, clear communication must be achieved, which can be tricky in an online classroom. However, there are ways in which communication—what Panos (2005) calls "emotional bandwidth," or trust via electronic mediums—can be established (p. 839). Specifically, Panos suggests using videoconferencing tools to establish trust, as the students involved can gain a better sense of the partners they are working with throughout the projects.

To further facilitate communication and encourage emotional bandwidth, I asked students to create wikis and blogs to write collaboratively with other students during group projects and to provide progress reports for the nonprofit and the instructor. In addition, the class created a wiki that served as a "library." For instance, in order to write the proposal, students had to research Muslims and Latinos to learn about these cultures. Each time a student found a relevant source, he or she would critique and annotate the source before adding it to the class wiki. This activity specifically promoted the programmatic outcome of Critical Thinking, Reading, and Writing.

By integrating wikis and blogs, I was also able to address an important objective of the program, Composing in Digital Environments, which is the fifth outcome developed by the council of Writing Program Administrators (the programmatic outcomes at ASU are based on the WPA outcomes). Although this outcome is not officially a programmatic outcome, as Maid and D'Angelo (2013) suggest, before graduating, students should be "comfortable working within both print and digital environments" (p. 13). Indeed, experience with digital writing is of growing importance in the field of technical communication, with Carliner (2009) suggesting that technical communicators are now expected to see a document through various stages, from inception, to production, and even publishing, thanks to the advent of open-source software (pp. 21–50). However, I also wanted students to consider the ethics of designing documents using digital media, which can be quite different than that of print-based text; essentially, I wanted students to become technologically literate, not just feel comfortable with composing in digital spaces. For instance, Gurak (2005) notes that the fast-paced nature of the Internet encourages short, to-the-point writing, leading authors to sometimes omit important details, which poses potential risks for the audience receiving the document (p. 210). Through reflection assignments, students in my course considered the values associated with digital writing, specifically regarding how an audience will interact with the information in the digital text and whether or not they are receiving the information they need.

ENHANCED COURSE OBJECTIVES AND THE GOALS OF SERVICE-ELEARNING

While meeting ASU's programmatic outcomes was the first priority, the online nature of the course and the service-learning element demanded the inclusion of additional course objectives. I structured my course around four objectives based on pedagogical values of service-eLearning as described by Dailey-Hebert and Donnelli (2010): Global Connectedness, Application of Skills, a Nonlinear Learning Environment, and Peer Learning (pp. 221–223). These values serve as guiding principles of service-eLearning, or the intended goals of the pedagogy. Although these objectives are not necessarily online-specific, because of budget cuts across the board, they may not be met in traditional learning environments. For instance, an f2f teacher with little access to online training may not have time to incorporate a supplemental Blackboard site, which would encourage nonlinear learning. Or an f2f class without technology may never give students the opportunity to communicate via technological mediums (i.e., videoconferencing software), and students may not learn the important skills of peer learning in an online environment, or digital communication, for the future. Thus, the online element proves to be successful in preparing our students to enter the workforce, even when resources might be scarce. In the following subsections, I describe the ways I have integrated these objectives within my service-eLearning course as well as the projects and/or assignments that students created to meet these objectives and how they benefit student learning.

Global Connectedness

Global Connectedness, the first objective, insists that students develop a personal and vested interest in global issues. This principle adheres to the guidelines set forth by the National Leadership Council Report (AAC&U, 2007) entitled *College Learning for the New Global Century*, which states that in their college studies, students must prepare for 21st century challenges by gaining personal and social responsibility on a local and global level. This responsibility includes attaining civic and intercultural knowledge and competence that are "anchored through active involvement with diverse communities and real-world challenges" (p. 2). Scholars in technical communication have suggested the discipline is responsible for preparing students to enter the workplace with citizenship and integrity, not just knowledge and skill (Staples, 1999, pp. 153–164). Dubinsky (2002) claims that students can develop this integrity by learning to work with community partners, recognizing their responsibility to continue that work after they leave academia (p. 63).

Starke-Meyerring (2005) reminds us that "community" is no longer synonymous with "local," as many technical communicators will work in transnational organizations and within global virtual teams (p. 470). As such, I often try to choose service partners with far-reaching efforts. For instance, as I mentioned previously, my students worked with the Peace Frame Initiative; the organization's primary efforts include creating videos displaying the commonalities between all cultures. For their final project, students developed similar videos that the organization could showcase at fundraisers and functions across the globe, helping various audiences recognize the stereotypes associated with these groups and understand the commonalities everyone shares, regardless of ethnicity. Working with such a diverse nonprofit to eradicate racism can help students learn of problems beyond their community and feel as if they have a stake in the endeavor, even from behind the confines of a computer. Furthermore, working with a diverse nonprofit helped students gain a sense of the programmatic outcome of Critical Thinking, Reading, and Writing, which encourages students to "understand the relationships among language, knowledge, and power including social, cultural, historical, and economic issues." Discussions of stereotypes help to ensure that students are gaining a sense of other cultures while learning to aid their community through their writing.

Application of Skills

If Global Connectedness is the development of cross-cultural understanding, then application, the second objective, is the ability to apply that understanding to the creation of real-world documents (Dailey-Hebert & Donnelli-Sallee, 2010, p. 222). As Thatcher (2009) argues, it becomes increasingly imperative for technical communicators to attend to the norms and rules of various cultures when producing a document (p. 169); this attention becomes even more important when designing an open-source Internet document that may be viewed by a larger, more diverse audience than a text-based project that is intended for in-house distribution within a specific organization. Johnson, Pimentel, and Pimentel (2008) claim that traditionally, technical communicators have been concerned with making information clear and accessible, often overlooking how an audience member of another culture might perceive their writing (p. 212). Indeed, writing for various audiences is an important outcome of technical communication and falls under ASU's programmatic outcome of Rhetorical Knowledge, which requires students to respond to the needs of various audiences while simultaneously adopting an appropriate voice, tone, and level of formality.

To promote greater audience awareness, Thrush (1993) suggests having students read articles that discuss variations in writing and document design within certain cultures (p. 272). Throughout my course, students were required to read articles such as Kostelnick's "Cultural Adaptation and Information Design: Two Contrasting Views" (1995), and they participated in an asynchronous discussion regarding how various cultures might perceive their writing in different ways. At the end of the course, students reflected upon how their writing addressed these differences, enabling them to further understand how their technological skills have the potential to benefit others when they consider alternative cultural perspectives.

Nonlinear Learning

The third objective, promoting a Nonlinear Learning Environment, helps students learn self-direction. The online learning environment characteristic of service-eLearning pedagogy promotes an indirect path to learning; students have deadlines, but they learn to manage their time around their work schedules. In part, the online platform allows for greater flexibility of learning, as students draw on knowledge from many resources, including the vast amount of information on the Internet, the teacher, and their peers. At the same time, they can use these resources to master difficult topics, replaying instructional videos and rereading archived discussion boards. Learning to access and retrieve information and apply that information to writing situations is an important aspect of the outcome of Knowledge of Conventions.

In addition, the principle of nonlinear learning aligns closely with the outcome of Processes, which includes the ability to learn to balance the advantages of relying on others with the responsibility of doing their part. For example, Muthusamy, Wheeler, and Simmons (2005) note the increase of self-managed work teams, or collaborative work groups who are not only responsible for the work they are asked to perform but also for monitoring their own performance (p. 53). Rather than being managed by a supervisor, these teams are responsible for gathering information, making decisions, and meeting the organizational goals. Group work for an actual organization mimics the work performed in self-managed work environments. In my course, the group project helped to prepare students to function within the self-managed work teams that are dominating the workplace.

Peer Learning

Collaborative group work can also encourage the objective of Peer Learning, which also corresponds with the programmatic outcome of Processes, as mentioned above. Peer learning and collaboration have been touted as goals of technical communication for decades (Rehling, 2005; Stratton, 1989); however, more recently, the conversation has turned to collaborating and participating in globally distributed work, with many writers and team members housed in organizations at various locations around the world. The removal of f2f communication in service-eLearning courses challenges students to interact with peers on a new level, insisting that they pay attention to verbal cues and inflections in tone, especially when videoconferencing. Collaborating via videoconferencing software demands that students become more in tune to their audience when using interactive software, learning to notice nonverbal cues such as eye movement, gestures, and even facial expressions (Hill & Harris, 2008, p. 79). Students in my course were required to "meet" via videoconferencing with the instructor at least once throughout the semester. I also integrated at least one synchronous discussion board a week so that students could ask for guidance from either the director or the instructor. In addition, students could meet with the director or me as often as they needed. Such communication is important to establish emotional bandwidth and a reciprocal relationship between all parties involved.

PEDAGOGICAL PERSPECTIVES: IMPLEMENTING SERVICE-ELEARNING IN VARIOUS PLATFORMS

Scholars of service-eLearning claim that the pedagogy can be utilized in a variety of classroom formats, including f2f, hybrid, and completely online. To encourage technological literacy in all classes, integrating service-eLearning within f2f or hybrid classes may be the easiest approach for instructors, especially those who have not taught in a fully online format before. In fact, Stoeker, Hilgendorf, and Tryon (2008) posit that simply creating open-source documents can enhance the service project, promoting deeper analysis by allowing increased accessibility and collaboration between multiple authors (p. 40). Many f2f instructors wanting to try service-eLearning can start by integrating technology into the classroom, asking students to develop multimodal projects for service partners. Or students can use blogs to collaborate in teams and keep progress logs of their work that the instructor can also access. Other ideas include asking students to write online, collaborative reflections regarding how technology impacted their document production and communication with other participants. Whether teaching f2f or hybrid courses, the Web-enhanced space can increase communication through software and foster greater collaboration among students and service partners.

Lastly, service-eLearning can be successful in fully online courses. Classes that are taught completely online pose their own set of challenges, in both the design of the course and in facilitating communication. While the online platform can increase the frequency of communication between all parties involved (instructor, service partner, and students), it may be difficult for students who are working to be able to meet at certain times. Synchronous and asynchronous meetings wherein discussions are transcribed and can be read at a later date by those unable to attend may mitigate scheduling concerns. I have found that using collaborative software with recording capabilities can be helpful. For instance, software like Spreecast can be used to create videos of interactive sessions, allowing students who could not attend the meeting access to the conversation. In addition, this software allows users to upload and share digital documents, including blogs and wikis, which can be beneficial when students are using such tools in the classroom. No matter what software is used, the instructor and students must learn to consider what technology will provide the greatest benefit to the project. It is the consideration of technology and the willingness to try new, unfamiliar software that will help the students be successful outside academia.

Regardless of course platform, instructors should incorporate reflection practices in order to connect course outcomes to the service project. Stemming from Dewey's (1938) belief that reflection serves as a natural conduit between theory and application, reflection is often used as a way to have students reflect on their work and the outcomes of the course, connecting these ideas to their sense of civic responsibility. As Bringle and Hatcher (1999) observe in "Reflection in Service-Learning: Making Meaning of Experience," reflection has "enormous potential for learning to broaden and deepen along academic, social, moral, personal, and civic dimensions" (p. 39). Scholars indicate that reflection in online formats can be deeper because students are able to return to archived discussions and reflect on prior conversations; the nonlinear format allows them to easily revisit information posted by peers and the instructor. In addition, online discussions can offer a way to participate in collaborative reflection with the service partner, which can give students a sense of how their work was received and the impact they made on their local community or world.

Reflection can be even more important when shifting a traditional f2f service-learning course to the online environment. For instance, I mentioned previously that I transitioned the online course described from an f2f course I taught in which students wrote similar documents and implemented fundraisers for the nonprofits. In the f2f course, students were able to see their efforts in action and interact with the nonprofit, viewing firsthand the directors' satisfaction with their work. Obviously, the online course had to be different—students could no longer actually hold the fundraisers. Some students found this element difficult; they couldn't understand how they were helping the nonprofit. In subsequent classes I've taught, I incorporated additional reflection assignments—a group reflection via discussion board and a live chat—wherein students can talk about challenges and offer advice for future students. In addition, during these final reflections, the directors of the nonprofits can join in and offer their advice and thoughts on the feasibility of the fundraisers, ideas for going forward, and their overall satisfaction with the project. After participating in these reflections, students write their own final reflections, and many have commented that the voices of the directors at the end of the course helped them recognize the importance of their work, regardless of whether or not they were able to physically help the nonprofit with the fundraisers. Thus, reflections can prove to be an integral component of the online service-eLearning course, especially when attempting to shift an f2f pedagogy online.

RESOURCEFULNESS IN AN AGE OF AUSTERITY: WHEN PROGRAMMATIC SUPPORT IS SCARCE

According to Donnelli-Sallee and Dailey-Hebert (2008), instructors who utilize the pedagogy can achieve greater success when supported by entire departments or programs within a specific department (p. 125). Because service-learning has merged with eLearning to form a new model, resources must be available for both, and faculty will need support teaching online and perhaps in forming relationships with community-service partners. Programs will need to develop a list of partners who are willing to work with the university. However, most departments probably don't have the resources or the funds for such support. When I first started teaching service-learning courses, I searched the Internet looking for nonprofits to work with and made calls to find out if any agencies might be interested in working with my students. This is one way teachers can find service partners when resources are not available. Over time, the more a teacher implements service-eLearning into his or her classrooms, relationships will form, and long-lasting partnerships will be established. Service-learning should be reciprocal in nature, meaning both partners and students should benefit from the projects; when reciprocity is repeatedly the outcome of a service class, more organizations will be willing to participate.

More importantly, technical communication programs will need to provide online training. Training is not only necessary for teachers who are required to shift their curriculum to a fully online format but also for those wanting to enhance their curriculum by adding an online component. At the University of

New Mexico, where I currently work, the entire college is shifting to Blackboard as the common learning management system. Fortunately, the university has offered training classes for teachers and students, as well as one-on-one training sessions for assistance developing course shells and curriculum. If structured training is simply not possible, there are tutorials instructors can turn to via Youtube videos, or the website www.lynda.com (which offers a 1-month free trial) can be an invaluable tool for learning platforms such as WebCT or Blackboard or software such as iMovie or InDesign. I also suggest letting the students have a voice in what technological tools they want to use. The new generation of students is accustomed to working with and using technology in their everyday lives and is typically more open to learning new programs. Letting students have a voice insists they consider the best software for document design and communication, which ultimately enhances the critical thinking skills that will transfer regardless of what situations they find themselves in within the workplace.

Aside from resources, instructors also need support for professional development and scholarship (Donnelli-Sallee & Dailey-Hebert, 2008, p. 125). Even when there is training for online teaching, there probably is little training for the service-learning element. This means we must continue to expand our efforts and publish, providing models and frameworks for others wanting to test the waters. Technical communication currently has numerous models of successful service-learning practices to turn to when designing curriculum; however, when it comes to shifting the pedagogy online, the scholarship is nonexistent. I have found success in publishing by simply receiving human-subject approval for almost all of the online courses I've taught in the past, then writing about the results of the classes from the perspective of the instructor and the students. For example, in an article in *Technical Communication Quarterly* (2014), I offer a small-scale assessment of a similar service-eLearning course, using student comments to determine their learning of course outcomes. If the field continues to evolve into the online arena, which certainly seems to be the trend, instructors need more models of successful practices to guide them. With the lack of departmental support that is so commonplace in universities across the country, it is up to us, as educators, to publish these models and guidelines.

CONCLUSION

Service-eLearning, although sometimes a challenge to implement, can be beneficial to technical communication as a field experiencing many changes. In the article "E-Service-Learning: The Evolution of Service-Learning to Engage a Growing Online Student Population," Waldner, McGorry, and Widener claim (2010) that a structured curriculum poses the greatest opportunity for a successful service-eLearning endeavor (pp. 123–150). The authors remind instructors to not just use technology for technology's sake, but instead, to think about how technology can benefit classroom practices. Unlike the generations before us, this new group of students interacts with technology every day. This generation prefers to multitask and participate in learning opportunities that use technology to engage them; as such, our curriculum needs to reflect these preferences (Dailey-Hebert & Donnelli, 2010, p. 219). As technical communication programs continue to offer more online classes, we must meet the demands of the institution and our students at the same time. Service-eLearning can be a way to meet these demands and still provide all students, distance education and traditional alike, with a quality education, which includes the ability to work within multicultural teams, to meet organizational expectations, and to adapt to the changing nature of technology within their chosen profession.

Just like we ask of our students, we as instructors should be reflecting on our work, improving our curriculum according to our own assessment of pedagogy within our classes. We should be publishing these results, encouraging collaboration and faculty development at our own institutions and others. And as hard as it may be, we need to embrace the challenges our field faces. In a classroom where I recently gave a talk on how to use technology to facilitate learning, the technology I was using ironically failed. Not to be deterred, I kept going with the lesson and improvised, still able to lead a fruitful discussion. Afterward, a student told me that my calmness under fire was perhaps the best lesson for her; I didn't waiver or show frustration, and that meant more to her than the discussion of technology. This is the example we must set for our students. Just like us, students can feel unprepared to meet the future of tomorrow. However, only when we learn to embrace the changes will our field continue to flourish and thrive in an age of austerity.

APPENDIX

Program Outcomes for ASU:

Rhetorical Knowledge

- Identify, articulate, and focus on a defined purpose
- Respond to the need of the appropriate audience
- Respond appropriately to different rhetorical situations
- Use conventions of format and structure appropriate to the rhetorical situation
- Adopt appropriate voice, tone, and level of formality
- Understand how each genre helps to shape writing and how readers respond to it
- Write in multiple genres
- Use appropriate technologies to organize, present, and communicate information to address a range of audiences, purposes, and genres

Critical Thinking, Reading, and Writing

- Use information, writing, and reading for inquiry, learning, thinking, and communicating
- Understand the relationships among language, knowledge, and power, including social, cultural, historical, and economic issues related to information, writing, and technology

Processes

- Be aware that it usually takes multiple drafts to create and complete a successful text
- Learn to critique their own and others' works
- Learn to balance the advantages of relying on others with the responsibility of doing their part

Knowledge of Conventions

- Learn common formats for different genres
- Learn standard tools for accessing and retrieving information
- Learn and apply appropriate standards, laws, policies, and accepted practices for the use of a variety of technologies
- Develop knowledge of genre conventions ranging from structure and paragraphing to tone and mechanics
- Control such surface features as syntax, grammar, punctuation, and spelling

REFERENCES

American Association of Colleges and Universities (AAC&U). (2007). *College learning for the new global century: A report from the national leadership council for liberal education and America's promise.* Washington, DC: AAC&U.

Ball, C. (2012). Assessing scholarly multimedia: A rhetorical genre studies approach. *Technical Communication Quarterly, 21,* 61–77.

Bourelle, T. (2012). Bridging the gap between the technical communication classroom and the internship: Teaching social consciousness and real-world writing. *Journal of Technical Writing and Communication, 42,* 183–197.

Bourelle, T. (2014). Adapting service-learning into the online technical communication classroom: A framework and model. *Technical Communication Quarterly, 23,* 247–264.

Bringle, R., & Hatcher, J. (1995). A service-learning curriculum for faculty. *Michigan Journal of Community Service-Learning, 2,* 112–122.

Bringle, R., & Hatcher, J. (1999). Reflection in service-learning: Making meaning of experience. *Educational Horizons, 77,* 179–185.

Carliner, S. (2009). Computers and technical communication in the 21st century. In R. Spilka (Ed.), *Digital literacy for technical communication: 21s century theory and practice* (pp. 21–50). New York, NY: Routledge.

Council of Writing Program Administrators (CWPA). (1999). WPA outcomes statement for first-year composition. *WPA: Writing Program Administration, 23*, 59–66.

Dailey-Hebert, A., & Donnelli, E. (2010). Service-elearning: Educating today's learners for an unscripted future. *International Journal of Organizational Analysis, 18,* 216–227.

Dailey-Hebert, A., Donnelli-Sallee, E., & DiPadova-Stocks, L. (2008). Service-elearning as an integrated pedagogy: An introduction. In A. Dailey-Hebert, E. Donnelli-Sallee, & L. DiPadova-Stocks (Eds.), *Service-elearning: Educating for citizenship* (pp. 1–8). Charlotte, NC: Information Age.

Dewey, J. (1938). *Experience and education.* New York, NY: Macmillan.

Donnelli-Sallee, E., & Dailey-Hebert, A. (2008). Service-elearning best practices: Possibilities for engagement. In A. Dailey-Hebert, E. Donnelli-Sallee, & L. DiPadova-Stocks (Eds.), *Service-elearning: Educating for citizenship* (pp. 119–130). Charlotte, NC: Information Age.

Dubinsky, J. (2002). Service-learning as a path to virtue: The ideal orator in professional communication. *Michigan Journal of Community Service-Learning, 8*, 61–75.

Eisner, S. (2010). A grave new world? Workplace skills for today's college graduates. *American Journal of Business Education, 3,* 27–50.

Gurak, L. (2005). Ethics in technical communication in a digital age. In C. Lipson & M. Day (Eds.), *Technical communication and the World Wide Web* (pp. 209–221). Mahwah, NJ: Lawrence Erlbaum.

Hill, S., & Harris, C. (2008). Service-elearning and professional writing. In A. Dailey-Hebert, E. Donnelli-Sallee, & L. DiPadova-Stocks (Eds.), *Service-elearning: Educating for citizenship* (pp. 75–86). Charlotte, NC: Information Age.

Hoover, K., Casile, M., & Hanke, R. (2008). How discussion boards drive course concept mastery in service-elearning. In A. Dailey-Hebert, E. Donnelli-Sallee, & L. DiPadova-Stocks (Eds.), *Service-elearning: Educating for citizenship* (pp. 59–74). Charlotte, NC: Information Age.

Huckin, T. (1997). Technical writing and community service. *Journal of Business and Technical Communication, 11*, 49–59.

Johnson, J., Pimentel, O., & Pimentel, C. (2008). Writing New Mexico white: A critical analysis of early representations of New Mexico in technical writing. *Journal of Business and Technical Communication, 22*, 211–236.

Kahn, H. et al. (2008). Creating international, multidisciplinary, service-elearning experiences. In A. Dailey-Hebert, E. Donnelli-Sallee, & L. DiPadova-Stocks (Eds.), *Service-elearning: Educating for citizenship* (pp. 95–108). Charlotte, NC: Information Age.

Kostelnick, C. (1995). Cultural adaptation and information design: Two contrasting views. *IEEE Transactions on Professional Communication, 38,* 182–196.

Maid, B., & D'Angelo, B. J. (2013). What do you do when the ground beneath your feet shifts? In K. Cargile Cook & K. Grant-Davie (Eds.), *Online education 2.0: Evolving, adapting, and reinventing online technical communication* (pp. 11–24). Amityville, NY: Baywood.

Manion, C., & Selfe, R. (2012). Sharing an assessment ecology: Digital media, wikis, and the social work of knowledge. *Technical Communication Quarterly, 21,* 25–45.

Morrison, J. et al. (1970). Roadhouse blues. *Morrison Hotel* [Record album]. Los Angeles, CA: Elektra.

Muthusamy, S., Wheeler, J., & Simmons, B. (2005). Self-managing work teams: Enhancing organizational innovativeness. *Organizational Development Journal, 23,* 53–66.

Panos, P. (2005). A model for using videoconferencing technology to support international social work field practicum students. *International Social Work, 48,* 834–841.

Parsad, B., & Lewis, L. (2008). Distance education at degree-granting postsecondary institutions: 2006–07 (NCES 2009–044). *National Center for Education Statistics, Institute of Education Sciences.* Washington, DC: U.S. Department of Education.

Rehling, L. (2000). Doing good while doing well: Service-learning internships. *Business Communication Quarterly, 63,* 77–85.

Rehling, L. (2005). Teaching in a high-tech conference room: Academic adaptations and workplace simulations. *Journal of Business and Technical Communication, 19,* 98–113.

Sapp, D., & Crabtree, R. (2002). A laboratory in citizenship: Service-learning in the technical communication classroom. *Technical Communication Quarterly, 11,* 411–431.

Scott, J. B. (2006). *Critical power tools: Technical communication and cultural studies.* Albany, NY: State University of New York Press.

Staples, K. (1999). Technical communication from 1950–1998: Where are we now? *Technical Communication Quarterly, 8,* 153–164.

Starke-Meyerring, D. (2005). Meeting the challenges of globalization: A framework for global literacies in professional communication programs. *Journal of Business and Technical Communication, 19,* 468–499.

Stoeker, R., Hilgendorf, A., & Tryon, E. (2008). Incorporating technology in service-learning: A case study of appropriate use. In A. Dailey-Hebert, E. Donnelli-Sallee, & L. DiPadova-Stocks (Eds.), *Service-elearning: Educating for citizenship* (pp. 31–44). Charlotte, NC: Information Age.

Strait, J. (2008). Constructing experiential learning for online courses: Two models of service-elearning. In A. Dailey-Hebert, E. Donnelli-Sallee, & L. DiPadova-Stocks (Eds.), *Service-elearning: Educating for citizenship* (pp. 75–86). Charlotte, NC: Information Age.

Stratton, C. (1989). Collaborative writing in the workplace. *IEEE Transactions on Professional Communication, 32,* 178–182.

Thatcher, B. (2009). Understanding digital literacy across cultures. In R. Spilka (Ed.), *Digital literacy for technical communication: 21st century theory and practice* (pp. 169–198). New York, NY: Routledge.

Thrush, E. (1993). Bridging the gaps: Technical communication in an international and multicultural society. *Technical Communication Quarterly, 2,* 271–283.

Waldner, L., McGorry, S., & Widener, M. (2010). Extreme e-service-learning (XE-SL): E-service-learning in the 100% online course. *MERLOT Journal of Online Learning and Teaching, 6,* 839–851.

Waldner, L., McGorry, S., & Widener, M. (2012). E-service-learning: The evolution of service-learning to engage a growing online student population. *Journal of Higher Education Outreach and Engagement, 16,* 123–150.

Walsh, L. (2010). Constructive Interference: Wikis and service-learning in the technical communication classroom. *Technical Communication Quarterly, 19*, 184–211.

Yang, Y. (2011). Learner interpretations of shared space in multilateral English blogging. *Language Learning and Technology, 15*, 122–146.

CHAPTER 7

Balancing Standardized Web-Based Pedagogy With Instructor Autonomy in Technical Writing Courses

Dirk Remley

On many campuses, the push to increase online course offerings goes hand in hand with the shrinking budgets that are part of a new era of austerity. For institutions and program administrators, Web-based delivery promises to eliminate the overhead costs related to classroom technology and maintenance, attract new student populations, and generate new revenues. Furthermore, content management systems make it easier than ever to share materials or reproduce a single class across multiple sections of a service course in business or technical writing. Standardization of online service courses offers a way to ensure students enrolled in various sections of the same course have consistent experiences across all sections. Further, standardization allows administrators to assign more sections to contingent instructors—part-time, graduate teaching assistant, and so on—secure in the knowledge that despite differences in approach to pedagogy and teaching experience, instructors will be delivering a consistent product.

Neither the shift to online courses generally nor the move to standardize courses is simple or uncomplicated, however. Moving several courses in a single program online can make for efficient curriculum design, but course developers must take care when designing specific pedagogies for a given course. While a single, multisection course will have specific content and objectives across sections, individual instructors might effectively use a wide range of pedagogical strategies to help students arrive at those programmatic goals. Consequently, in any course standardization initiative, there is a delicate balance to be struck between the programmatic or institutional desire for consistency and efficiency and individual teachers' desire for the autonomy to integrate assignments and materials with which they have had success.

This chapter explores the balancing act that takes place—or should take place—whenever a program is considering developing a "master course" for a technical writing service program, especially in time of austerity. I argue that an effective approach balances institutional or administrative desires for standardized pedagogy with instructors' desire for individual autonomy. Striking a balance can be a daunting prospect, but it is not impossible. My insights grow out of my own experiences as a member of the development team that designed standardized service courses for business writing and technical writing at Kent State University in 2012. I focus on the strategies we used to implement a successful standardized technical communication course at Kent State: (a) taking a team approach to development of a master course; (b) making rigorous use of best practices for online course design; and (c) creating policies to ensure consistency between sections yet allow for teacher autonomy. By describing our processes and results, as well as reflecting on the lessons learned, I hope to provide both practical and theoretically grounded advice for others who find themselves—of their own accord or as the result of institutional fiat—navigating the standardization process.

In this chapter, I begin by briefly discussing the context for our technical communication standardization initiative at Kent State. I will describe the development of our transdisciplinary course design team before I discuss the ways in which best practices shaped our master courses. I will then discuss the important role policy plays in bridging the gap between autonomy and consistency. I conclude by considering the relationship between the drive to standardize all sorts of courses and Taylorism.

LOCAL CONTEXT: PUTTING COURSES ONLINE AT KENT STATE

Our local context at Kent State is likely to feel familiar to technical communication faculty at other medium-to-large state-funded schools. According to its website, Kent State University has an enrollment of over 42,000 students across the multicampus system (2012). As such, Kent State is the second-largest state-supported institution in Ohio. Some 80% of students receive some kind of financial aid (Kent State, 2012). Like most state-supported institutions nationally, the university has seen support from the state decline each year since the recession of 2006 began. By 2010, annual budget shortfalls, an institutional desire to increase enrollments and course profitability, and increasing demand from international students had triggered a campus-wide initiative to offer more classes online (Beck, 2011; Costello, 2011; DKS Editors, 2011; Moore, 2011; Wilkinson, 2011).

While the technical writing and business writing courses had been offered online prior to the 2010 university initiative, it was in this context—growing demand for online service courses and a general push toward more online courses across campus—that directors of the Writing Program decided to standardize course designs for the service courses in business and technical writing. To meet the demand for online offerings, the department needed more instructors who could teach online, but many had never taught Web-based courses. Therefore, the impetus to standardize the Web-based courses was the need to facilitate instruction of the courses by less experienced faculty as enrollments rose.

TAKING A TEAM APPROACH

One of the first questions anyone considering standardizing courses must answer is, Who will design the course? Once the decision to standardize the services courses had been made, the Writing Program administrators approached the Office of Distance Learning and Continuing Education (ODCE) for assistance. ODCE helped fund the effort, paying the design teams for their time and labor. As a funded project, the University would own the course templates and be able to control some of the design. The design, however, would be a team effort.

Assembling the right people to design the master course is a critical first step in the process of standardizing courses. The perspectives and experiences of team members can mitigate or exacerbate some of the most common problems with the development of online learning spaces. For example, as I reported (2013), if the designer(s) of a standardized course has/have limited experience with certain pedagogies or tools or have negative experiences with particular tools, then the pedagogy they design will discourage use of those tools for a given purpose or activity, or the activity may be redefined because of the designer's experience with the tool. The Committee on Best Practices in Online Writing Instruction found that 20% of respondents acknowledged inheriting a course designed in a CMS by someone else (CCCC, 2011). Further, 58.6% who taught at 4-year universities, similar to my institution, inherited a course designed by someone else (Hewett et al., 2011, p. 19). The study also found that, for the most part, course designers are content specialists.

To guard against the sort of curricular/pedagogical tunnel vision described above, the Writing Program Committee selected faculty for the design team who had deep experience with the course and with teaching online. Teams consisted of two non–tenure track faculty, one tenure-track or tenured faculty member, and at least one instructional technologist who was well acquainted with the primary delivery tool that would be used for the master course—Blackboard Learn. Because individual members had previously developed and successfully applied Web-based pedagogies, the team could draw on a diverse range of designs as inspiration for the template.

There was a sense of equality within the teams amongst the content specialists due to our shared experience with the CMS and course generally. No one "pulled rank" to influence a decision. Different stakeholders naturally started the process with different ideas about what made for a good online class. The administration wanted a single "master course" template to standardize the pedagogy across sections, which caused much consternation among members of the design team. The teams debated internally about how to design a "master course" template, consistent with administration's concern about uniformity of instruction, that also allowed faculty the latitude to integrate pedagogies with which they were more familiar and had demonstrated success.

We found a way to bridge the gap by focusing on best practices for online pedagogy and by implementing a policy that supported both the administrative goal of consistent experiences for students and the design team goal of flexibility for teachers. We were able to arrive at a design that would be considered good quality pedagogy by the Quality Matters program. Quality Matters is a national program that assesses the pedagogical soundness of Web-based courses at several institutions, serving as a primary certification program to ascertain pedagogical soundness. Indeed, the Ohio Board of Regents encourages institutions to use it as the standard of quality assurance for online courses (Reed, 2012). In the sections that follow, I will offer some highlights of our application of best practices and then return to a discussion of policy.

THE CHALLENGES OF STANDARDIZING PEDAGOGY

Before our design team could begin thinking about best practices, we gave careful consideration to the pros and cons of teaching and learning with standard CMS technologies. Course management systems' template features facilitate easy sharing of materials and courses; however, these platforms pose two general challenges for master course developers—one for instructors, one for students—both centered on proficiency with technology. The technology makes it quite easy to standardize by simply developing a course template that can then be transferred between instructors and copied into new sections. However, such standardization assumes that instructors and students are proficient—and equally so—with the tools required to use the pedagogy built into the template. The tools have the potential, however, to create a hurdle that can negatively affect instruction and learning.

As I have reported (2013), administrators and faculty need to understand to what degree CMS tools can and should be used to standardize writing pedagogy. Generally, standardized pedagogy that is facilitated through technology assumes equal access to and proficiency in the technology among instructors as well as students. When that proficiency is lacking, the technology's transparency is eliminated and adjustments accounting for those differences are needed. Further, students' performance can be negatively affected if they cannot use the tools easily. The pedagogical experiences of designers of standardized Web-based courses may negatively affect course design and pedagogy, and students may need training to use certain tools that are frequently used as part of the CMS-facilitated, Web-based experience so that their performance will not be negatively impacted by their inability to use such tools.

Some older scholarship recognizes that Web-based learning assumes that students have certain proficiency with the tools used for instruction that they may not actually have. Morgan, for example, found that many students experienced problems using WebCT tools for learning, including access, proficiency with technology skills, and student learning preferences (2003, p. 50).

Course Management Systems and Standardization

Design teams have considerable research to draw on as they develop their thinking about course standardization in their local context. This scholarship includes a number of arguments for and against use of CMS tools in pedagogy generally and with respect to standardization. In addition to administrative and teacher perspectives, these also include student's perspectives. I review some of these here.

Students like knowing (a) that any given section of a particular course will offer the same content, assignments, and workload, and (b) that the course has been designed with a particular objective in mind and that all sections will provide instruction toward that objective (Fotz, O'Hara, & Wise, 2004; Hara, 2010). A common complaint of students taking writing classes is that another section of the same course involved much less (or more) activities than their own such that students in their or the other section had a much different learning experience.

Teachers like standardization because a predesigned course minimizes preparation time, and such a design tends to integrate a "best practices" model. Design teams can develop a template that integrates many attributes of best teaching practices for a given course—course content, facilitating interaction among students, and between student and teacher. Faculty who have research responsibilities or teach a range of different courses appreciate a streamlined approach to teaching (Fotz et al., 2004; Hara, 2010). The less work they have to do to prepare a class, the better. In addition, if graduate students and adjunct faculty teach a course, someone can train them to use a particular pedagogy, not having to worry that a style they decide to use might fail. This attribute also makes standardization of such courses (those taught primarily by graduate students and adjunct faculty) appealing to administrators. However, this argument assumes that designers understand appropriate course content and how to facilitate learning of that content.

While these benefits are acknowledged, many seasoned teachers enjoy the freedom to teach a class in a way that best suits their teaching style or that they believe helps students to learn course content and skills better than others they may have tried (Nelson, 2003; Samuels, 2004, 2007).

A Standard Course Design to Facilitate Customized Technical Writing Pedagogy

In addition to the proficiency with course tools, our design team considered how to bring the classroom experience into the Web-based environment. It is generally understood that Web-based courses involve a much different experience than any classroom-based experience for a variety of reasons, not the least of which include students being more responsible for their own learning and course material needing to be more accessible electronically than the instructor merely making it available and talking about it in class. A number of scholars have reported on differences between classroom-based writing pedagogy and Web-based pedagogy (Blythe, 2001; Cargile Cook & Grant-Davie, 2005; De Pew, Fishman, Ruetnick, & Romberger, 2006; Harrington, Rickly, & Day, 2000). Recently, the NCTE-sponsored Committee on Best Practices in Online Writing Instruction reported on a survey of the ways instructors tend to use Web tools to facilitate writing instruction; these vary, based on the instructor's experiences with specific technologies, further calling attention to the importance of building autonomy into course design. Therefore, a primary concern in developing Web-based courses is to make sure that one is not merely moving their classroom pedagogy onto the Web. I review some of that material below.

Generally, the more structure there is to a course design, and the better able students are to recognize and understand that structure, the more students are able to use the course site effectively to learn with it. Further, any instructional materials must provide comprehensive information so students fully understand concepts and what is expected of them in the course and on specific assignments. In addition, instructors should encourage electronic interactions between students and with the instructor to facilitate active discussion and engagement with the course.

BEST PRACTICES

Once we had worked through some of the issues related to standardization and course technologies, we looked to "best practices" as a guide for developing effective course shells. A growing body of scholarship has considered ways that tools used to facilitate Web-based pedagogies affect learning (see, for example, Cargile Cook & Grant-Davie, 2005 or De Pew et al., 2006). The Quality Matters Program (2011) assesses the effectiveness of online course design using a rubric that also uses many of the attributes identified by Chickering and Ehrmann (1996). The eight items they review are

- Course Overview and Introduction
- Learning Objectives (Competencies)
- Assessment and Measurement
- Instructional Materials
- Learner Interaction and Engagement
- Course Technology
- Learner Support
- Accessibility

DISCUSSION OF BEST PRACTICES CRITERIA IN COURSE DESIGN

In this section, I describe how we integrated particular attributes of the QM criteria into the course design. These include facilitating lively participation and discussion, opportunities to practice writing and receive detailed feedback on it, encouragement for interaction with the instructor and other students, clear expectations for assignments and instructional materials that allow for diversity in learning approaches.

Discussions

Course design must provide spaces and prompts that encourage discussion amongst students and between student and teacher. Students learn more when they are able to interact with others in discussions about topics than if they never engage with anyone as they work on assignments and read course materials. For a technical writing course, whole-class discussions can revolve around certain exercises as students discuss how they responded to a given task. The variety of possible forms of writing, also, facilitates participation in class discussions (Gray, 2002, p. 37). The Quality Matters rubric (2011) encourages faculty to invite students to introduce themselves to the entire class the first week of the term of a course with an informal posting to a discussion board. This builds a sense of community as students share some interesting information such as their major, work background, and interests. Gray also encourages draft postings for peer review within the writing process (2002).

Course Expectations

Students also need to understand what the teacher expects relative to all the work associated with the course: participation, readings, and assignments. General information about what instructors expect of students may be provided explicitly in the course syllabus; or, if there is a list of specific expectations, one can provide that list as a separate file. Instructors can also post a message to a discussion area at the beginning of each week about what students should be working on that week. That helps students understand weekly expectations.

While one may be able to acknowledge verbally to students their expectations with a given assignment in a classroom setting as they set up the assignment or respond to questions, this information must be much more explicitly provided in a Web-based setting. With each activity assigned, whether it is an exercise or graded assignment, in addition to conveying the task, one needs to provide specific information about what course goals/objectives are met with the task, how students should go about addressing the task, and what criteria will be applied to the submission to determine a grade. The more students understand the task, how it fits in with the course, and what they need to do to do well on it, the better they can do without having to ask many questions because of ambiguities an instructor may not perceive.

Master Course Design

In the technical writing master course, there are 14 different discussion topics provided. Most weeks involve a single topic; however, some involve two topics. These range from students introducing themselves to discussing topics such as effective phrasing in instructions and choosing appropriate graphics to include in instructions. I have been able to add some instructional materials within my own sections of the course to facilitate additional instruction with particular attributes of technical writing I tend to highlight, such as Web design and integration of graphics.

Most of the instructional materials include links to websites and Word files. There are a few videos provided as well. We also included specific information about particular course objectives associated with assignment prompts to help students understand the relevance of a given assignment. Finally, we include a rubric so students can see on what they will be assessed and how certain attributes will affect grading.

To facilitate work on some collaborative assignments, I eliminated some of these general discussions. Students work in the small-group discussion space, and that workshopping is considered in their participation grade for the course.

CONTENT

Content should not change because of delivery mode or across sections because of instructor; students taking any course are expected to learn the same skills whether the course is delivered in a classroom setting or via the Web. Typically, courses are designed for a curriculum with specific objectives in mind; while there may be some variance in specific ways to reach those objectives, the objectives remain consistent across sections. Generally, the explicit goal of introductory technical writing courses is for students to learn how to provide specialized information to engage various audience—some with expertise in a given subject and others who have no such expertise—toward a particular purpose.

Master Course Design

The design teams for both the business writing and the technical writing courses observed that we all used certain assignments for each course. In particular, these included a project proposal related to a final project, a project that simulated a writing task in a workplace setting, and activities associated with developing visuals. Consequently, the master course includes instructional materials related to all of these assignments and activities. However, there is enough flexibility that one could include additional assignments and activities geared toward their own pedagogy.

For example, in the technical writing course, we included assignments in which students develop graphics and visuals to include within different genres, and I modified the assignments within my own sections. One assignment designed for the master course specified a bifold brochure/pamphlet; however, I integrated a website-development requirement with that assignment. Figure 1 is the assignment display in the course site.

Item number 2 was designed into the master course, while I added item number 1.

Another assignment that I modified from the one provided with the master course was one related to developing instructions, another common assignment in technical communication courses. The assignment in the master course asks students to prepare instructions for a task with which they were familiar, letting them pick the task. I modified that to include two specific tasks: depositing/withdrawing funds to/from an ATM and jump starting a dead vehicle battery. The types of assignments are uniform, but instructors can customize specific items related to them.

BALANCING STANDARDIZATION AND AUTONOMY: THE ROLE OF POLICY

Our design team was confident we'd built a class that represented best practices, but it took an explicit policy statement to give us confidence that we had achieved our goals for teacher autonomy. Beyond the work of a design team, program committees can develop policies to encourage some autonomy for instructors, as ours did. The Writing Program Committee (WPC) developed a policy statement that invites individual faculty to modify the course toward their own pedagogy. However, it also ensures that such modifications are reviewed and consistent with sound pedagogical practices. This policy is part of a larger policy document; however, the portion relevant to this chapter follows:

> These templates serve as default course designs for Writing Program instructors who may not have had previous experience with teaching online. The WPC will revisit and update these templates as necessary. Faculty

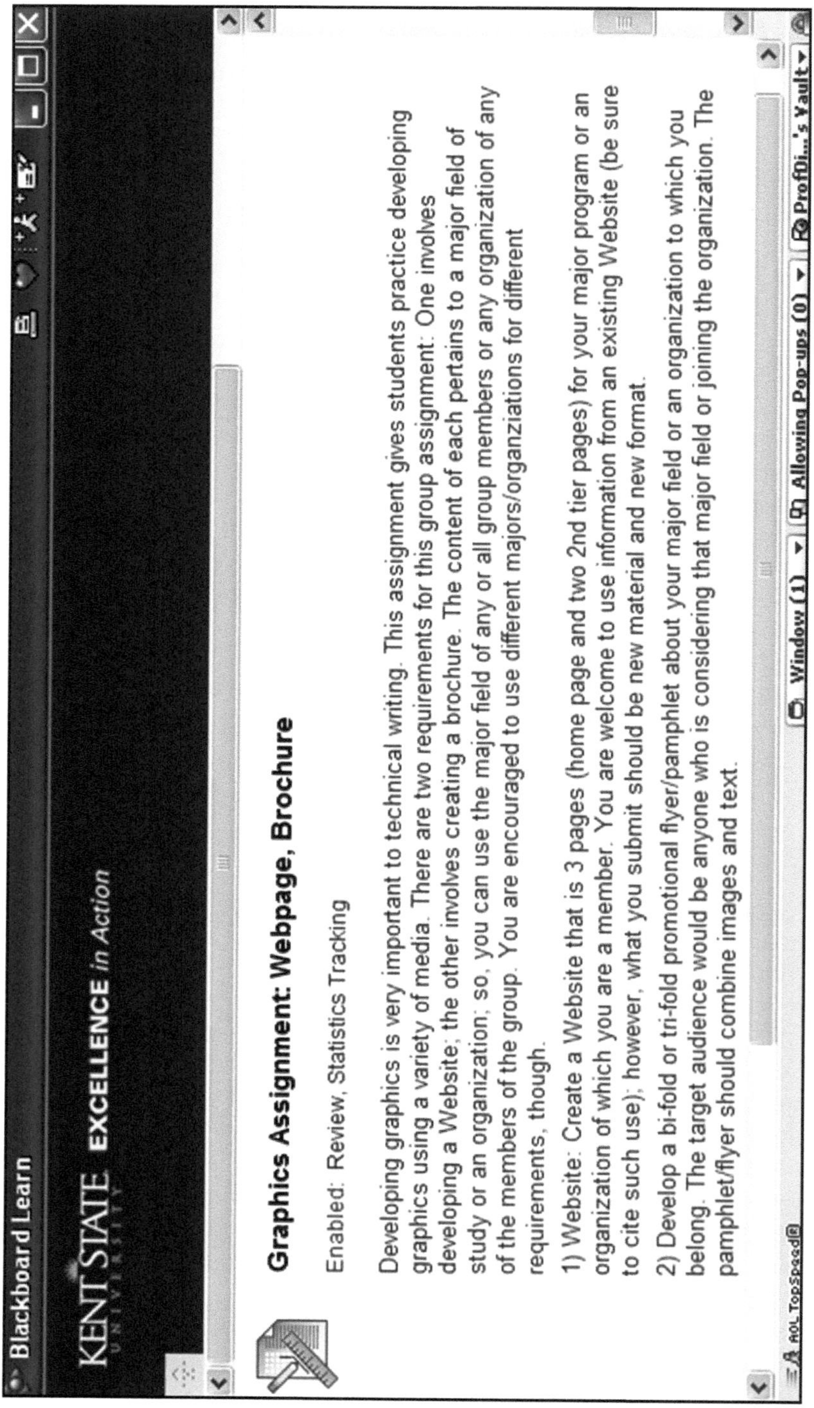

Figure 1. Assignment display for graphics assignment.

> deemed qualified to teach DL courses by the WPC may alter, modify, or discard the templates in favor of their own courses and designs, as long as these are approved by the WPC and meet the quality online standards that the university has adopted for online course design, as considered appropriate for a writing course by the WPC. While different course structures are allowed, unnecessary duplication should be avoided.

The policy also encourages faculty who have never taught the course before but who are deemed qualified to teach it to use the template. However, the committee also will ask one with more experience to act as a mentor for that instructor for the semester in which he or she first teaches the course. This ensures that the less experienced instructor learns the substance behind the instructional materials provided in the site.

In addition, at the end of the semester, instructors who have taught any Web-based sections of the course submit a reflection to facilitate assessment. This reflection is a narrative that allows the instructor to indicate observations about challenges students experienced related to course design and/or technologies.

IMPLICATIONS OF STANDARDIZATION OF WRITING PEDAGOGY: TAYLORISM

The standardization described in this chapter reflects a newer approach to pedagogical design than most veteran faculty teaching technical writing courses have experienced. Many of us have been able to develop our own course around a particular pedagogy that we experienced as students or that has evolved for us as we have read scholarship and tinkered with various approaches in our own classes. A dynamic that cannot be helped in standardization of pedagogy is the potential for a design team's work to be deemed as absolute pedagogy while others who teach the course are viewed as laborers simply delivering the designed course pedagogy and whose work involves "merely" responding to student questions and grading assignments and other assessments. Our teams were able to address some concerns related to maintaining instructor autonomy, even on a limited basis, while bringing uniformity to courses; however, the potential for negative attributes of Taylorism to impact writing pedagogy is evident.

The best outcome of such standardization is the synergy built into the pedagogy. It is developed through a team effort, and the team includes veteran instructors who have been practicing sound pedagogy for years. As "experts," the team members design a course template that integrates "best practices" generally. By encouraging instructors to use it while adding to and/or deleting from it toward customizing for their own pedagogy, the course design begins with a good pedagogical approach. Instructors are not likely to change the design dramatically while customizing it for their own approach and style.

However, the more instructors are discouraged from customizing for their own pedagogy, or simply accept the design as the best practice and let the design teach the course, the more such standardization simulates Taylorism. An important attribute of Taylorism is that it reduces the need for highly trained workers to be hired while improving efficiencies (Taylor, 1913). This enables graduate students and less trained part-time faculty to teach the course. Recall that these were the primary motivations in the development of the master courses for the writing program—so graduate students and part-time faculty who did not have much experience with teaching the course could teach it.

CONCLUSION

Austerity measures are forcing technical communication programs to facilitate more courses online while providing quality instruction and consistent content across sections. Using design teams that include seasoned faculty with Web-based teaching experience and recognizing the value of integrating instructor autonomy in pedagogy will facilitate good pedagogy in this environment. Expert faculty, whether tenure-track or non-tenure-track, can identify assignments appropriate to any course while allowing for flexibility with those assignments. However, faculty status cannot drive design or design team membership; expertise with the pedagogical tools and materials as well as an understanding of best practices must drive design. The same can be said for the need to allow for some autonomy; instructors with certain technical skills should be allowed to use those to help their students learn, while those with less advanced technical skills may be able to use other tools to facilitate similar learning toward a course learning outcome.

Further, a mentoring program for less experienced faculty ensures that the design does not teach itself, minimizing the potential for Tayloristic dynamics to emerge. This allows graduate students and adjunct faculty opportunities to teach some sections, lowering instructional costs while ensuring instructional quality. It also enables those less trained faculty to gain quality training along the way.

REFERENCES

Beck, C. (2011, December 8). University pushing to offer more online classes. *kentwired.com*. Retrieved from http://www.kentwired.com/latest_updates/article_47cc617e-e4e6-555c-b145-d2dbb3e9ea06.html

Blythe, S. (2001). Designing online courses: User-centered practices. *Computers and Composition, 18*, 329–346.

Cargile Cook, K., & Grant-Davie, K. (Eds.). (2005). *Online education: Global questions, local answers*. Amytiville, NY: Baywood.

CCCC Committee on Best Practice in Online Writing Instruction (OWI). (2011). *Survey of fully online distance-based instruction.* Retrieved from http://www.zoomerang.com/Shared/SharedResultsSurveyResultsPage.aspx?ID=L246LPBS94AZ

Chickering, A. W., & Ehrmann, S. C. (1996). Implementing the seven principles: Technology as lever. *AAHE Bulletin* 3-6. Retrieved from http://www.tltgroup.org/programs/seven.html

Costello, J. (2011, January 30). University hopes to triple online courses within the next 3 years. *kentwired.com.* Retrieved from http://www.kentwired.com/latest_updates/article_1890482c-7364-5ad4-a238-66f64c1bc8b3.html

De Pew, K., Fishman, T., Ruetenik, B. F., & Romberger, J. (2006). Designing efficiencies: The parallel narratives of distance education and composition studies. *Computers and Composition, 23*, 49–67.

DKS Editors. (2011, February 2). Our view: Distance learning tuition needs to change. *kentwired.com.* Retrieved from http://www.kentwired.com/opinion/article_fadc4c62-90f6-5c0b-9e78-e35fcc156bf1.html

Foltz, C. B., O'Hara, M. T., & Wise, H. (2004). Standardizing the MIS course: Benefits and pitfalls. *Campus-Wide Information Systems, 21*, 163–169.

Gray, R. (2002). Assessing students written projects. *New Directions for Teaching and Learning, 91*, 37–42.

Hara, B. (2010, November 18). Academic freedom versus mandated course content. *The Chronicle of Higher Education.* Retrieved from http://chronicle.com/blogs/profhacker/academic-freedom-vs-mandated-course-content/28764

Harrington, S., Rickly, R., & Day, M. (2000). *The online writing classroom.* Cresskill, NJ: Hampton.

Hewett, B. K., Minter, D., Gibson, K., Meloncon, L., Oswal, S., Olsen, L., et al. (2011). *Initial report of the CCCC Committee on Best Practice in Online Writing Instruction (OWI).* Retrieved from http://www.ncte.org/library/NCTEFiles/Groups/CCCC/Committees/OWI_State-of-Art_Report_April_2011.pdf

Kent State University. (2012). About. *Kent State University.* Retrieved from http://www.kent.edu/about/facts/StudentBody.cfm

Moore, D. (2011, October 4). University budget grows after state cuts funding. *kentwired.com.* Retrieved from http://www.kentwired.com/latest_updates/article_b15a1411-ad9d-508e-8b3f-ace7e4d61b9e.html

Morgan, G. (2003). Faculty use of course management systems. *ECAR Educause center for applied research.* Retrieved from http://www.educause.edu/ir/library/pdf/ers0302/rs/ers0302w.pdf

Nelson, J. L. (2003). Academic freedom, institutional integrity and teacher education. *Teacher Education Quarterly, 30*, 65–72.

Quality Matters Program. (2011). *Quality Matters Standards rubric 2011–2013.* Retrieved from http://www.qmprogram.org/files/QM_Standards_2011-2013.pdf

Reed, T. (2012). Online learning growing, ESC promoting quality matters. *The Intercom.* Retrieved from http://intercom.escco.org/2011-2012/11June/QualityMatters.html

Remley, D. (2013). Templated pedagogy: Factors affecting standardized writing pedagogy with online course management systems. *Writing and Pedagogy, 5*, 105–120.

Samuels, R. (2004). The future threat to computers and composition: Nontenured instructors, intellectual property, and distance education. *Computers and Composition, 21*, 63–71.

Samuels, R. (2007). Revealing codes: Cognitive mapping of writing, computers, and grants at the postmodern university. *Kairos, 7*(3). Retrieved from Kairos.technorhetoric.net/7.3/coverwebb/Samuels/kairoswriting.html

Taylor, F. W. (1913). *The principles of scientific management.* New York, NY/London, UK: Harper & Brothers. Retrieved from http://books.google.com/books?id=HoJMAAAAYAAJ&pg=PA3#v=onepage&q&f=false

Wilkinson, M. (2011, September 12). More KSU students resort to online courses. *kentwired.com.* Retrieved from http://www.kentwired.com/latest_updates/article_3014bbab-40c7-52a7-81fb-04ae42f58d29.html

CHAPTER 8

Working Conditions, Austerity, and Faculty Development in Technical Writing Programs

Ed Nagelhout, Denise Tillery, and Julie Staggers

Faculty development is an important part of the way we are responding to austerity at our institution. We administer writing programs at a university where literally all writing courses—particularly our business and technical writing service courses—are taught by part-time instructors, both graduate students and adjuncts (Tillery & Nagelhout, 2013, p. 28). Even before the onset of the Great Recession, our teachers worked under horrid conditions. Salaries were meager, healthcare benefits were inadequate and unstable, prospects for leveraging even "full-time" part-time teaching into permanent teaching positions were practically nonexistent. In addition, the limited material and intellectual support for adjunct instructors—from their densely packed offices with shared desks and phones in a ghetto of trailers far from the English Department to a departmental culture that had translated "intellectual property" and "academic freedom" into "don't share anything"—meant instructors had to do things on their own and alone. In our program, as elsewhere, adjunct instructors were operating from deep within the terrain of austerity long before the economic collapse of 2008.

The recession hit the state of Nevada particularly hard, making an already dire situation much worse (Hebel, 2010; Kelderman, 2011). The state's economy is heavily dependent on gambling and tourism, but state revenue from those sources dropped nearly 25% between 2007 and 2010. The downturn in tourism ripped through the local economy, intersecting with the national mortgage crisis to disastrous local effect. After 20 years as the nation's fastest-growing city, Las Vegas in 2008 saw the simultaneous implosion of its three leading industries: gaming/tourism, construction, and real estate. By fall 2011, unemployment had reached 13%, and Nevada was leading the nation in home foreclosures. The state legislature cut funding for higher education by 30% between 2009 and 2011. However, the higher education funding formula, which channels monies generated at UNLV to the state's land-grant university in Reno and to the community colleges that serve Nevada's few and far flung rural students, ensured that cuts fell disproportionately on UNLV. Students saw tuition increases of up to 60% in a period of just a few short years. By the fall of 2011, UNLV had cut 400 jobs—nearly 20% of its total staff—mostly in administrative and support services. The title of an article in the *Chronicle of Higher Education* in September of 2011 captured the austerity zeitgeist on our campus: "In Nevada Harsh Reality Hits Higher Education: Years of Budget Cuts Sap Campuses and Morale" (Kelderman, 2011). The 2011 legislative budget cycle was particularly damaging. The legislature held higher education budget cuts to 15%, but the budget process left the campus community deeply scarred. Plans had been afoot to use "financial exigency" to void tenure (when our professional writing faculty received electronic pink slips in May of 2011, all but one of us were tenured), and UNLV made the national media for several days running after a proposal to eliminate the entire Philosophy Department went public. Large-scale reductions of administrative and support staff, unit reorganizations and hiring freezes that left academic departments threadbare and faculty on overload, the elimination of merit pay, and the conversion of what had been fairly decent health benefits to a catastrophic coverage plan left tenured faculty angry, dismayed, and generally demoralized. But tenured faculty were protected during the downturn, even on our campus, in ways that other university employees—including adjunct instructors—were not. Indeed, one consequence of faculty hiring freezes and cuts to support staff is an increasing reliance on adjunct instructors, who now are doing their jobs with even less ad hoc support from departmental support staff and with even worse health benefits. In addition to the adverse conditions described above, the recession and subsequent austerity measures impose three particular burdens on our teachers:

- In an attempt to spread out fewer courses (with larger enrollment, naturally), our program limited instructors to three sections per semester, a less than full-time load (although UNLV does offer benefits for instructors teaching at least three courses). The burden is on us as administrators to make those three sections equal *less* than a full-time job, enabling instructors to pick up other job shifts if necessary.

- UNLV has been increasingly recruiting international students, who pay higher tuition rates. These international students are concentrated in the business college and enter into the Business Writing course with typical English-language-learner challenges. Enrollment caps for the online sections of Business Writing were increased by five, significantly increasing the workload for teachers in the sections most popular with international students.
- Wages have been cut across the board, and more significantly, health benefits have been drastically reduced. Essentially, these measures punish long-term teachers for their loyalty by drastically decreasing take-home pay for people who have been in these positions for years.

As program administrators making peace with the new normal, it is up to us to alleviate or mitigate these burdens on our hard-working instructors in whatever way we can. We have two broad, underlying commitments for our writing programs:

1. *Coherence and Consistency for All Courses in Our Writing Programs*
   This means that every student in every section of a writing course will have basically the same experience: All students are expected to meet the same outcomes, begin with the same primary sources and materials, perform the same set of tasks and activities, and be held to the same standards for evaluation, whether it's a single course with multiple sections or a multicourse, multisection program like our professional writing minor.
2. *Work-Life Balance for All Teachers in Our Writing Programs*
   In order to achieve #1, we must provide the kind of consistent support for teachers that helps increase their skills and makes them more efficient and effective inside and outside the classroom. From our position within the institutional hierarchy, there is little we can do to alleviate meager salaries, inadequate healthcare benefits, or lack of career stability for our teachers, but we can work to ensure they have a manageable work-life balance.

In this chapter, we interrogate the working conditions for writing teachers in higher education; the ways austerity has only exacerbated these working conditions for, especially, the most marginalized teachers; and the important role faculty development plays for all technical writing programs. Our goal is to redefine working conditions, teacher expectations, and teacher success from the perspective of the teachers working in a writing program rather than from the perspective of the work that needs to be done in a program. In other words, expectations for success and "work" can be adjusted in order to create a more equitable work-life balance.

WORKING CONDITIONS, TRADITIONAL WORKLOAD EXPECTATIONS, AND AUSTERITY

Typically, working conditions for teachers in higher education are defined by such things as the physical environment, privileges gained for working in a particular environment, autonomy, a voice in program governance, or the potential for stress. For writing teachers specifically, these working conditions might include class size, teaching loads, the availability of teaching materials, adequate space for conferring with students individually, adequate time for responding to and evaluating student work, and access to additional resources that may enhance classroom instruction.

Piled on the very notion of working conditions is the time-honored image of the writing instructor. The typical image is of an instructor who spends long hours working at home in isolation, constructing brilliant lectures on mundane points of grammar, and engaging in conversations with students that include pithy comments on both arcane and battle-tested composition techniques. Further, she (for she is almost inevitably female) sits for more long hours in cramped offices, mentoring individual students to improve each sentence of an essay, and lugs giant stacks of student papers home every weekend, writing evaluations on student drafts that are characterized by a multitude of marginal comments and lengthy end comments. All of these activities are meant to enable students to go boldly into the world to deliver well-crafted prose to the masses with grace and aplomb.

We are all familiar with these mythical images and the exaggerations they imply. They have shown up implicitly and explicitly for decades in discussions of the roles writing teachers are cast in: mother (Schell, 1998); Delphic Oracle (Lauer, 1976); writing coach (Hamilton, 1997); heroic liberator (Freire, 1984); to name a few. These discussions have all attempted to promote the value inherent in writing teachers working so many hours without ever attempting to undermine this masochistic approach to writing instruction. Almost all of the roles described in our literature assume that writing teachers are working more hours than they should. The idealism of overwork appears to be the default. And, sadly, we cultivate this image on an almost daily basis through our stories, our practices, and the awards we bestow on the "best" teachers.

For example, the following post is nearly 5 years old, but similar sentiments arise on discussion boards and blog posts on a near-weekly basis:

> First, I suggest, we not expect more of FYC instructors; in fact, our in-house workload report of two years ago indicates we're overdue on expecting less. It finds that when we try to follow best practices, we are trying to work—and work well—for 70+ hours/week. (Baker, 2009)

This post on the Writing Program Administrators Listserv claims that when a composition program tried to follow best practices, their teachers were working 70+ hours per week. Now, this post does not mention if these instructors are teaching a "full load," but a speaker at a recent Conference on College Composition and Communication (CCCC) made the claim that if a teacher of composition taught three sections, then that teacher was automatically working overtime. Why is that? More importantly, if this kind of workload comes from "best" practices, then why don't we reevaluate our "best" practices? Best practices should at least be manageable.

Overcoming these "best practices" is made even more difficult, especially for smaller or marginalized writing programs, like most technical writing programs, in light of the primarily miscast ideals of indoctrination and surveillance promoted by position statements from such bodies as the NCTE, CCCC, and TYCA. These documents, though well-intentioned and not without merit, generally fail to consider the reality of teaching writing, especially for a part-time instructor, an adjunct, or a graduate student teacher.

Too often, these documents offer idealistic visions of "best practices," with often unmanageable expectations. For example, the NCTE's Two-Year College Association for years offered "Guidelines for the Academic Preparation of Two-Year College English Faculty" in the form of a list of "characteristics" of effective 2-year college English faculty:

- Reflective practitioners and flexible teacher/scholars
- Classrooms are student-centered
- Understand diversity and teaching to diverse ethnic, economic, and ability populations
- Challenge their students
- Teaching is grounded in theory and research
- Collaborate with colleagues in developing curriculum
- Actively serve their colleges and their communities
- Participate in professional community through conferences, presentations, publications, and ongoing scholarship (NCTE, 2004)

While this list may prove helpful to the diligent search committee (or the taskmaster program administrator), the very idea of "characteristics" seems to imply innate talents rather than learned skills developed over time and within particular contexts. In any professional environment, the universal ideal of "characteristics" innate to the best (or best prepared) writing teachers seems antithetical to many of the aims that we promote as educators and writing teachers. The majority of teachers, especially writing teachers, work hard at developing their teaching skills. Teaching writing requires *lifelong* learning.

Likewise, the Conference on College Composition and Communication has an equally daunting list of expectations for the "Preparation and Professional Development of Teachers of Writing" (CCCC, 1982) (and this list is just the first level):

> I. Programs for the preparation and continuing education of teachers of English and language arts, at all levels, should include opportunities for prospective and active teachers:
> - to write
> - to read and respond to the writings of students, classmates, and colleagues
> - to become perceptive readers of our own writing, so that we can ask questions about, clarify, and reshape what we are trying to express
> - to study and teach writing as a process
> - to experience writing as a way of learning which engages us in intellectual operations
> - to learn to assess the progress of individual writers by responding to complete pieces of their writing and studying changes in their writing
> - to study research and other scholarly work in the humanistic discipline of the teaching of writing
> - to study writing in relation to other disciplines (CCCC, 1982)

This list for preparing and continuing the education of teachers of writing seems particularly daunting in light of the 70+ hour work weeks required (by myth) of the best teachers in our profession because few (if any) of our contingent faculty will have been prepared to the standards described by the CCCC position statement or possess the innate "characteristics" described by the TYCA position statement when they begin teaching in our technical writing programs. Moreover, as a recent editorial in *Peer Review* lamented, "Many faculty are never sufficiently acculturated to the institution(s) where they teach" (Tritelli, 2002, p. 3). Traditional writing teacher support meant providing a syllabus, ordering a textbook, then leaving the teachers alone (in the name of academic freedom).

While we certainly don't wish to impugn the value of these various position statements, most program administrators will struggle to apply them in effective and efficient ways, especially at institutions where the oppression of contingent faculty appears both natural and expected. Since adjuncts and graduate students make up the vast majority of writing teachers across the country, working conditions for most writing programs, including technical writing programs, will not be improved by the dependence on contingent faculty to teach the majority of courses (Meloncon & England, 2011, p. 401).

Our profession must get past a set of ideals that are too often unattainable from a practical perspective. We must acknowledge the kind of life issues that make everything worthwhile, and these life issues are not more work, not more study, not more time spent with student papers. Writing teachers should be expected to be writing teachers. No more. No less. More importantly, we all have to define what that means (no more, no less) both locally and explicitly.

The oppression of contingent faculty is a stain on our profession. Despite much recent work to amend this problem (see the Special Issue of *College English* edited by Palmquist & Doe, 2011, in particular, or the NCTE College Section Working Group, 2011), we can still do more. As a discipline, we can talk all we want about unionizing and getting more money for our contingent faculty; these are noble, worthy aspirations. We need to pursue these aims, but we can also work from within to be sure that we are supporting our writing teachers. Department chairs and writing program administrators, especially, need to avoid being complicit in this oppression. While only midmanagement in the institutional hierarchy, they are in a unique position to address this problem through humane action.

Very few universities pay full-time salaries for writing teachers not on the tenure track, and so not only are writing teachers making a pittance, but they often have minimal training in the teaching of writing. They may have had some mentoring when they were graduate students, often a single course at the beginning of their careers, and they may be required to go to the occasional workshop, more often than not sponsored by a book publisher, but they are rarely given strategies for reducing their teaching workload, thereby allowing them to live a more fulfilling life or possibly get some sleep during the semester (see Brady 2001; Martin, 2000; McMahon & Green, 2008).

To begin to address the question of a manageable work-life balance, we define teacher workload as that "time" a program expects a teacher to commit to the activity of teaching. Reviewing teacher workload in terms of the "time" necessary to successfully perform the expected activities forces the program (or, by default, the program administrator) to define, locally and explicitly, the expectations for professional work. In our technical writing program, we define the teaching commitment in these ways:

- *Time in class* means those 3 hours per week that teachers are physically in the classroom. For online teachers, this would mean designating 3 hours per week physically available to students online, times when a teacher will be working on-site: interacting with students, offering "lectures," and dealing with questions or concerns.
- *Time in office hours* means those 3 hours per week that most departments require teachers to physically be in their office (or, for online sections, logged into the course management system). Obviously, time in office hours is flexible, and the efficient teacher uses this time to perform other tasks when no students visit.
- *Time preparing for class* means those hours that teachers use to get ready for class: organizing their classroom materials and handouts, arranging possible discussion topics, and constructing any lectures they might present.

- *Time responding to and evaluating student writing* means those hours that teachers spend with student drafts and final products. For us, responding to and evaluating are two different activities. Responding to student writing are those times when a teacher offers feedback in-process. Evaluating student writing are those times at the end of the process when a teacher grades a final product. This is an important distinction to make with both teachers and students.
- *Time in teacher development* means those hours that teachers spend improving the quality of their teaching and the quality of instruction in the program. In other words, time must be set aside every semester for professional development. This may include staff meetings, norming sessions, activity-specific workshops, course/program development consultations, program governance, and so on rather than overtime; it should be proactive rather than reactive; it should be a long-term consideration rather than a short-term response; and it must improve all aspects of the program environment.

Our teachers should be willing and able to teach effectively within the parameters of a "full-time" workload. No more. No less. While some may argue that "teaching is more than a job and less than a life," we all need to challenge the veracity of this statement (at least the first part). Teaching, we believe, is *not* more than a job. It is a remarkably satisfying job. It is an important job. But it is, at the end of the day, a job. And we all, at the end of the day, go home to a life. All of our thinking, all of our planning, all of our administrative energy should begin with an understanding of the "time" necessary to do the job well.

We conclude this section by repeating our earlier statement: Many (if not all) part-time instructors, adjuncts, and graduate student instructors have been operating from deep within the terrain of austerity, long before the economic collapse of 2008. But any form of austerity has the potential to put even more negative pressure on the working conditions of these marginalized instructors. To take just one example, many programs and departments faced with shrinking budgets and shrinking staffs at a time when many institutions are faced with increased student demand are increasing their online course offerings at breakneck speed. These courses offer a number of advantages for students and institutions alike, but for inexperienced online instructors or those developing online courses from scratch, the amount of invisible and uncompensated labor is significant. Austerity means we will need more faculty development, not less, in order to help teachers meet the challenges of doing more with less. Austerity does not necessarily lead to stagnation and should not dry up opportunities for teachers to learn and engage, but keeping these options open requires program administrators to think strategically.

FACULTY DEVELOPMENT, NOT PROFESSIONAL DEVELOPMENT

We believe instructors teaching a four-course load should have the kinds of support that allows them to do *all* of the work associated with those four classes in a 40-hour work week. And our instructors who have been limited to a three-course load should not let their work expand to take 40 hours. They need to protect their time to pursue other important activities in their lives, such as doing freelance writing, pursuing other part-time jobs (both of which may be necessary to make up for salary cuts), or caring for children. Our conversations with teachers about workload begin with the time that our program expects a teacher to commit to the *activity* of teaching. Writing teacher support must, primarily, address workload and time commitment, helping teachers meet the high (yet manageable) expectations established by the program. Materials and activities should be designed as part of the normal, expected workload. In other words, writing teacher support should not constitute "overtime" for teachers to be successful in our writing programs.

We want to begin by emphasizing the difference between what has traditionally been seen as "professional development" or "professionalization" and what we see as "faculty development." As we see it, professional development aims to help instructors find their way out of the forest of contingent teaching and into, if not tenure-track jobs, jobs that are at least "better"—in whatever way individual teachers define better—than their current jobs. Professional development is important for those seeking to navigate up through the academy. We offer both ad hoc and formal assistance with professional development—everything from workshops on teaching with technology to a job-search support group—for graduate students and part-time instructors preparing to enter the job market. We think it's fair to offer this sort of professional development as extracurricular. Individuals who perceive benefits from these sorts of opportunities can reasonably be expected to invest time and energy and shoehorn them into their schedule.

Professional development initiatives are important; however, they do little to improve the immediate circumstances or day-to-day lives of adjunct instructors, be they graduate students who hope to move on to tenure-track jobs after graduation or place-bound, long-term part-time instructors. Not everyone is looking to, or able to, move up and/or out. We contend that what adjunct instructors need extends beyond professional development to *faculty* development. By faculty development we mean the kinds of initial training, mentoring, material support, and ongoing skill development that help instructors become more effective and more efficient teachers. In addition, this sort of faculty development allows us to implement a comprehensive assessment program—which helps us assure coherence and consistency for students—without unduly burdening instructors.

Faculty development—whether it be mentoring, workshops, or norming sessions—has never been seen as important by either teachers or administrators in our department. Arguments of academic freedom, arguments of burden, and arguments of theory versus practice have historically undermined faculty development. We see these arguments as misunderstanding the motives and goals of faculty development, which include

1. Helping teachers be more effective
2. Helping teachers be more efficient
3. Helping teachers develop their strengths as teachers
4. Giving teachers responsibility for their work
5. Offering teachers choices

In 2007, we initiated a service course redesign that entailed development of an assessment-driven master course approach with face-to-face, hybrid, and fully online course models. Most of our adjuncts lacked training in rhetoric, composition, or professional writing. We quickly discovered we could not implement this sort of programmatic change without also addressing the working conditions of our instructors. We quickly came to see that faculty development is inextricably bound up with course development and that it is always situated in particular, local contexts.

There may be some broad, global strategies for improving the work-life balance of instructors on the margins that cut across institutions and disciplines. Certainly, in the next section, we attempt to extrapolate some issues that cut across all institutions or types of institutions. However, we believe strongly that the power to enact change almost always lies at the local level. Indeed, what is needed and what counts as faculty development is likely to be highly situated and context-specific. Thus far, we have been focused on our own institutional context; faculty development might look quite different in other contexts.

Faculty development allows us to improve working conditions by creating manageable expectations for students and faculty. Faculty development does not mean dictating activities or prescriptive approaches to teaching. Faculty development does not hinder academic freedom. Instead, faculty development means that teachers have conversations about expectations, about standards, about definitions for successful writing, articulating the kinds of support structures necessary for their long-term success. More importantly, faculty development interrogates and defines working conditions in the context of a writing program. When we talk about faculty development with faculty, we establish where the working conditions are the most oppressive. Faculty development is more than making the writing classroom better. Faculty development is about making our *lives* better.

In our writing programs, we define a full-time workload as 10 hours per week, which means that a trained teacher should never be expected to spend *more* than 10 hours per week on any one section of a course. This includes time in class, time in office hours, time preparing for class, time responding to and evaluating student writing, and time in professional development. If a full-time workload is not stated explicitly, along with the attendant activities, it's too easy for adjuncts, part-time instructors, and graduate student teachers, in particular, to take on unmanageable burdens, even if only inadvertently or through inexperience. The goal, then, in reviewing writing teacher workload is to establish standards for successful teaching as part of a manageable work-life balance.

To improve program quality, we must improve working conditions in terms of an accepted normal workload. In our writing programs, we use standard syllabi and common readings. We also distribute a workload calendar that maps out the expectations for work time spent each week of the semester, which is never more than 10 hours per week, even when responding to student writing. We make these expectations clear. If teachers are spending more than 10 hours per week on the course, they have to let us know so we can help them. Developing a proactive, integrated approach to faculty development can be a challenge. Two strategies we rely on are building faculty development into the expected workload and designing faculty development that saves teachers time.

A COMMITMENT TO A MANAGEABLE WORK-LIFE BALANCE

Faculty development is never an add-on to the regular work week. We accomplish this by integrating formal development sessions into the semester in strategic ways. Our faculty development meetings are held when the classroom workload is a bit lighter. This usually means those times between assignments, after one assignment has been returned, and before the next assignment is submitted, ensuring that an hour in a meeting won't be overtime. Within the meeting itself, we focus on strategies that will help teachers save time. For example, we had a 90-minute workshop designed to help teachers reduce their paper load. We discussed time-saving strategies, reviewed sample papers as a group, and put response to student writing in the context of larger writing program goals. After this workshop, teachers typically save at least 1–2 hours each time they respond to a set of papers. By our program math, a 90-minute workshop should save them 6–10 hours over the course of the semester.

An important first step in devising locally effective faculty development is to rethink the relationship between faculty development and working conditions. One of the standard bits of advice passed from WPA to WPA is that if you want to have successful faculty development, you either "feed them" or "pay them." We don't completely disagree with tips like these; snacks are always nice. However, we do believe suggestions like "feed them or pay them" arise from a perspective that local working conditions drive/constrain faculty development. If we accept this premise, then faculty development will always be a burden, always more work placed on already full plates. Rather than treating working conditions as antecedent to faculty development, we began treating faculty development as a vital element of our working conditions. From our perspective, if faculty development is a working condition, then it should be considered a legitimate part of teachers' everyday lives. We stopped looking at the ways in which working conditions affected faculty development (usually negatively) with the result that instructors had so little time left in their week that we were faced with either asking them to participate in faculty development with no compensation during their "time off" or forgoing faculty development altogether. We started looking at the ways in which effective faculty development might have a positive impact on working conditions. We reasoned that if we built faculty development into the work week, we could use those development opportunities to help teachers save time when they could and allow them to spend more energy on developing their teaching strengths. The key was devising faculty development that addressed the double problems of workload and time commitment.

A second important step in devising effective faculty development is rethinking teachers' workloads. Work weeks, workloads, and pay structures in academia broadly are dominated by a white-collar, salaried employee paradigm of both the worker and the work week. However, 75% of the people teaching in American university classrooms during any given week are not tenured or tenure-track faculty; they are graduate teaching assistants, adjunct, and non-tenure-track instructors. On average, they earn less than $3,000 per course, which breaks down to as a little as $17.00 per hour over the course of a 16-week semester (rates averaged from Curtis & Thornton, 2013, p. 9). Clearly, in terms of pay scale and working conditions, these teachers more closely fit the worker profile of blue collar or "hourly" workers.

We address working conditions from a variety of perspectives in higher education, but we rarely discuss working conditions as a matter of pay for hours worked. Across the academy, this certainly has much to do with the white collar/professional identity of the professoriate. In English departments particularly, we also struggle mightily with the mythos of the "good instructor." We may be paying adjunct instructors like hourly employees, but the collective soul of the English Department expects them to work like Ivy League lawyers trying to make partner before they turn 30.

Working conditions shape and are shaped by the ways that we think about our programs. Faculty development has the potential to affect working conditions: a response to management, worker, and product identities constructed by both blue-collar and white-collar ideologies (as described in particular by Berry, 2005; Bousquet, 2008; and Tokarczyk & Fay, 1993). When we made the commitment to a manageable work-life balance at UNLV, we were forced to approach faculty development from a different perspective—teacher work rather than program products. But we were determined to improve program quality and reduce teacher workload without increasing administrator workload (for our work-life balance was important, as well). To do this, we needed "trained teachers" and faculty development that served a variety of purposes.

To begin, we define "trained teacher" as someone who had participated in our mentor program for those who were teaching a particular course for the first time. This mentoring includes the syllabus, the course readings, assignment options, and other support materials at least 3 weeks before the beginning of the semester, along with regular (weekly) meetings during that first semester. Mentoring provides training—for subject matter and course technologies as well as for teaching strategies—for instructors without prior experience with the course, but it also provides a social support system that lasts beyond the semester. We do this as both support and training. One reason we do mentoring is to help guide new teachers through all aspects of the workload as effectively and as efficiently as possible so that they might meet these expectations without sacrificing the quality of the course or the quality of their teaching.

More importantly, for us, "trained teacher" also means we provide further (significant) support to assist all teachers in our writing programs to be more effective and more efficient teachers: regular (3–5 times per semester) faculty development meetings, such as reviewing program-specific course and classroom materials; norming assessment and evaluation materials; introducing program-specific processes or procedures; and providing an open forum for program governance. Each in-semester meeting has three components: (a) the state of the program (how is everyone doing?); (b) the future of the program (what does everyone need?); and (c) a focused activity (what can we do better?). This further provides an opportunity for governance of the program as well as a forum for input and insight.

Redefining writing teacher support forces us to examine best practices for our particular program and come to some agreements on the best ways to teach and assess the course(s) in our technical writing programs and the best ways to prepare our teachers to do that. This cannot be a top-down approach, however, but an ongoing conversation, a collaborative effort that provides a voice for all stakeholders in the program. Support should be about negotiating ideals, for the program, for the teachers, for the students, creating a "culture of support" (Marek, 2009, p. 275). Without these program-based writing teacher conversations, we really are just a bunch of individuals teaching the same course.

STRATEGIES FOR REIMAGINING FACULTY DEVELOPMENT

The writing program administration literature is full of advice on workshops and teacher training, but rarely does it focus on teacher training that makes the lives of our teachers better or improves the quality of our working environments. Rarely does it focus on teacher training designed to help teachers work less, but that is an explicit goal of our faculty development approach. How then do we improve working conditions based on a 10-hours-per-week definition of workload?

First, program expectations, and the professionalization process, must be defined within the parameters of a full-time workload. Reviewing writing teacher workload from this perspective is to establish attainable standards for successful teaching. Defining professional behavior in terms of the time necessary to perform the expected activities not only makes explicit program expectations for teachers, but can also help administrators recognize those activities and materials necessary for teachers to work more effectively and more efficiently. For example, to assist our writing teachers, we offer a sample 10-hour-per-week schedule (such as this one from our Business Writing program) (see Figure 1).

A sample time schedule helps teachers organize their time and provides a common language for helping them understand where they may need better support or individual assistance in meeting the expectations. More importantly, a sample time schedule provides an impetus for developing program materials and themes for faculty development meetings.

Second, reimagining faculty development based on 10 hours per week gives us permission to focus on local issues and trust in the long term, such as when we determined that our teachers were spending far too much time on student papers without seeing any significant benefits for students. This problem was exacerbated by increasing numbers of ESL students; we had several instructors who felt it was their responsibility to copyedit every single student's assignment and mark every single error. So we changed our approach. Our goals were to reduce teacher workload, reduce administrator workload, and improve student learning. To achieve these goals, we required teachers to respond to a draft in-process and then (after student revision) evaluate it as a final product. And each teacher was asked to make this process clear to students. When students submit a draft to a teacher, they know that they will be given higher-order feedback at this stage, designed to help them improve their draft and improve their writing. As a program, we make it clear to students that our teachers are writing teachers, they are not editors. The job of the writing teacher is not to respond to every single error; instead, the job of the writing teacher is to help students understand how to improve their writing by offering guidance where necessary, along with higher-order and lower-order strategies. The students are expected to develop their skills for revision/editing and are expected to take responsibility for their learning. We reimagined our administrator responsibilities by making program expectations for teachers and students more explicit and more transparent. Since we changed our approach, teachers spend less time with student papers, and student complaints have dropped significantly.

Third, our teacher development meetings must serve a variety of purposes. For example, we regularly use faculty development meetings to do collective norming. At least once per semester we choose one project and evaluate a sample

Business Writing at UNLV

10-Hour-Per-Week Schedule for Teachers

Class size of 24 and Standard Syllabus

Face-to-Face, Hybrid, Online (Classtime = time in class and/or time online)

Note: Office Hours should be used for Prep Time, Respond Time, or Evaluate Time

Pre-Semester: Maximum 6 Hours to set up each section on WebCampus & Prep Week 1; 2 Hours Professional Development Meeting

Week 1: 3 Hours Classtime; 7 Hours to Prep Week 2 & Week 3

Week 2: 3 Hours Classtime; 7 Hours to Prep Week 4 & Week 5

Week 3: 3 Hours Classtime; 2 Hours to Evaluate Exercises 1 & 2; 5 Hours to Prep Week 6

Week 4: 3 Hours Classtime; 6 Hours to Evaluate Exercises 3, 4, & 5; 1 Hour Professional Development Meeting

Week 5: 3 Hours Classtime; 4 Hours to Respond to Introductory Project Drafts; 3 Hours to Prep Week 7

Week 6: 3 Hours Classtime; 4 Hours to Evaluate Exercises 6 & 7; 3 Hours to Prep Week 8

Week 7: 3 Hours Classtime; 3 Hours to Evaluate Introductory Projects; 4 Hours to Prep Week 9 & Week 10

Week 8: 3 Hours Classtime; 4 Hours to Respond to Case Project Drafts; 3 Hours to Respond to Research Design Plan Drafts

Week 9: 3 Hours Classtime; 2 Hours to Evaluate Research Design Plans; 4 Hours to Prep Week 11 & Week 12; 1 Hour Professional Development Meeting

Week 10: 3 Hours Classtime; 3 Hours to Evaluate Case Projects; 2 Hours to Evaluate Exercise 8; 2 Hours to Prep Week 13

Week 11: 3 Hours Classtime; 4 Hours to Evaluate Exercises 9 and 10; 3 Hours to Evaluate Progress Reports

Week 12: 3 Hours Classtime; 7 Hours to Prep Week 14 & Week 15

Week 13: 3 Hours Classtime; 4 Hours to Respond to Job Project Drafts; 3 Hours Optional Time

Week 14: 3 Hours Classtime; 6 Hours Optional Time; 1 Hour Professional Development Meeting

Week 15: 3 Hours Classtime; 6 Hours to Respond to Final Report Drafts; 1 Hour Professional Development

Week 16: 4 Hours to Evaluate Job Projects; 4 Hours to Evaluate Final Reports and Final PAMs; 2 Hours Professional Development Meeting

Figure 1. 10-hour-per-week schedule for teachers.

paper as a staff. This provides an opportunity to discuss our evaluation criteria in more sophisticated ways, interrogate our grading practices, and come to some agreements on the most important considerations in grading that particular project. These open conversations also serve to encourage program assessment practices, wherein teachers are active participants in the process, with a voice in program development. Teachers in the program view assessment as program development, rather than surveillance, and they feel comfortable offering suggestions for not only improving their teaching, but also for improving the project, the course, and/or the program.

Similarly, our "Egg Timer" workshops help teachers focus on response or evaluation (but not both) in order to work with student writing more effectively and more efficiently. If we focus the workshop on responding to student writing, the whole staff looks at a sample of writing from a course to determine the most important problems present in the sample, based on the stated evaluation criteria for the assignment. We then respond in a way that would help the writer both revise the draft and improve their writing skills. And the teachers have 10 minutes to do it. Once the egg timer rings (or beeps if it is electronic), teachers have a conversation about what they determined were the most important problems with the draft and discuss strategies for offering comments as efficiently as possible. The goals of the meeting are to show teachers how much they can actually get done in 10 minutes and to help teachers develop good habits for responding to student writing without getting bogged down in minutiae or line editing, thereby reducing the time teachers spend in this activity. Similarly, this meeting also helps with consistency and coherence across sections because teachers are talking about the project, about student writing, about the importance of the stated criteria, about best kinds of feedback, and about the best strategies for doing their work. The very best part, of course, is the social support, an opportunity for conversation, a chance to collaborate and learn from each other. For instructors who are not on campus full time, who teach only at night or online, these opportunities for conversation are critically important.

Although professional development—as we defined it earlier, the extra support we provide to enable adjuncts to compete for other long-term career prospects—is not the primary goal of our faculty development sessions, it is one of the long-term outcomes. Our teachers think about assessment and information literacy in new ways and learn to articulate their choices in the context of program objectives, and these skills help them develop more sophisticated and competitive application materials.

CONCLUSION

Long before the Great Recession hardened into what increasingly looks like a permanent age of austerity—a new normal—instructors, most on the margins in our departments, were grappling—usually solo—with the edict that they do more with less, or worse, do everything with nothing. As programs and departments, as well as institutions and systems, grapple with diminishing budgets, smaller faculties, and increased student demand, adjunct instructors will play an increasingly important role in even small programs. A proactive and integrated approach to faculty development may be the best defense (or the smartest offense) for adjunct instructors and contingent faculty alike.

REFERENCES

Baker, M. L. (2009, January 27). *Best practices in hard times* [Message posted to WPA-L].

Berry, J. (2005). *Reclaiming the ivory tower: Organizing adjuncts to change higher education*. New York, NY: Monthly Review.

Bousquet, M. (2008). *How the university works: Higher education and the low-wage nation.* New York, NY: New York University Press.

Brady, J. L. (2001). Farewell to teaching. In M. Dubson (Ed.), *Ghosts in the classroom: Stories of college adjunct faculty—and the price we all pay*. Boston, MA: Camel's Back.

Conference on College Composition and Communication (CCCC). (1982). *CCCC position statement on the preparation and professional development of teachers of writing.* Retrieved from http://www.ncte.org/cccc/resources/positions/statementonprep

Curtis, J., & Thornton, S. (2013). Here's the news: The annual report on the economic status of the profession, 2012–13. *Academe,* 4–19. Retrieved from http://www.aaup.org/file/2012-13Economic-Status-Report.pdf

Freire, P. (1984). *The politics of education: Culture, power, and liberation.* Westport, CT: Bergin & Garvey.

Hamilton, S. J. (1997). *Collaborative learning: Teaching and learning in the arts, sciences, and professional schools* (2nd ed). Indianapolis, IN: IUPUI Center for Teaching and Learning.

Hebel, S. (2010, March 19). State cuts are pushing public colleges into peril. *Chronicle of Higher Education, 56*(27), A1–A22.

Kelderman, E. (2011, September 2). In Nevada, harsh reality hits higher education. *Chronicle of Higher Education, 58*(2), A1–A8.

Lauer, J. (1976). The teacher of writing. *College Composition and Communication, 27,* 341–343.

Marek, K. (2009). Learning to teach online: Creating a culture of support for faculty. *Journal of Education for Library and Information Science, 50,* 275-292.

Martin, R. (Ed.). (2000). *Chalk lines: The politics of work in the managed university.* Durham, NC: Duke University Press.

McMahon, D., & Green, A. (2008) Gender, contingent labor, and writing studies. *Academe, 94,* 16–19.

Meloncon, L., & England, P. (2011). The current status of contingent faculty in technical and professional communication. *College English, 73*, 396–408.

National Council of Teachers of English (NCTE). (2004, November 4). *Guidelines for the academic preparation of two-year college English faculty*. Retrieved from http://www.ncte.org/groups/tyca/featuredinfo/119956.htm

NCTE College Section Working Group on the Status and Working Conditions of Contingent Faculty. (2011). Statement on the status and working conditions of contingent faculty. *College English, 73*, 356–359.

Palmquist, M., & Doe, S. (Eds.). (2011). Special topic: Contingent faculty. *College English, 73.*

Schell, E. (1998). *Gypsy academics and mother-teachers.* Portsmouth, NH: Boynton-Cook.

Tillery, D., & Nagelhout, E. (2013). Theoretically grounded, practically enacted, and well behind the cutting edge: Writing course development within the constraints of a campus-wide course management system. In K. Cargile Cook & K. Grant-Davie (Eds.), *Online education 2.0: Evolving, adapting, and reinventing online technical communication* (pp. 25–44). Amityville, NY: Baywood.

Tokarczyk, M. M., & Fay, E. A. (1993). *Working-class women in the academy: Laborers in the knowledge factory.* Amherst, MA: University of Massachusetts Press.

Tritelli, D. (2002). From the editor. *Peer Review, 5*(1), 3.

SECTION THREE

External Challenges and Opportunities

CHAPTER 9

Googling Academe

Timothy D. Giles, Angela Crow, and Janice R. Walker

The prospect of trimming information technology (IT) costs and increasing network security would certainly delight many administrators faced with a state legislature demanding university budget cuts. Unfortunately, IT cost-cutting measures may create questions about intellectual property and privacy issues for students and faculty. In a technical communication program, IT resources intertwine with the curriculum, and if those resources erode, questions are raised about professionalizing students.

Georgia Southern University (2009) has decided to rely on Google Apps as a resource for faculty and students interested in creating websites. It remains unclear what the best curricular response should be to this set of decisions to move away from university-hosted websites. Should we limit instruction to resources supported by the university, pay out of our own pockets for an outside hosting service that provides the requisite resources for teaching, or should we ask the university to contract with an outside hosting source that is more robust than Google but costs more money? Issues at stake here include how university policies are shaped by constraints and how curriculum should be supported. To stay within the given institutional framework or to establish our own is a complicated decision.

This chapter explores these issues by focusing on a decision to reduce IT costs by outsourcing services and on the effect on one technical communication program. It begins by providing background on our specific professional and technical writing program, situates the issue within the larger disciplinary context, and concludes by describing the ways intellectual property, privacy, and curricular issues are affected.

BACKGROUND

To understand the relations to austerity at a local level, along with the university and the department's abilities to negotiate in the midst of austerity concerns, this section will overview the kinds of power framings available to the department. We started with a limited ability to negotiate for viable resources. Our department began at Georgia Southern University when the Department of English and Philosophy, which had over 60 full-time faculty members, was broken up. At the time, it was the largest department, and its size exceeded some colleges on our campus. The provost who enabled this change was a business professor whose area of expertise was in management and for whom the Department of English and Philosophy had become a management problem. What emerged in 1998 were the Department of Literature and Philosophy and the Department of Writing and Linguistics. In 2004, the University System of Georgia's Board of Regents officially granted our department the authority to offer a BA in writing and linguistics.

Within the Department of Writing and Linguistics, the areas of study that emerged, as they are currently named, are Professional and Technical Writing, Creative Writing, Writing Studies, and Linguistics. Though the degree has since been revised at the behest of a former Dean of the College of Liberal Arts and Social Sciences, we still maintain our department's four areas. They have since expanded, and the professional and technical writing area provides students working on a BS in IT an option for a second discipline, which allows them to take 18 hours of our courses. We also have designated a minor in professional and technical writing that is distinct from the minor in writing for other students. The former requires the professional and technical writing courses, while the latter allows any upper-division writing and linguistics courses.

The Online Bachelor of General Studies (BGS) has been important to the Professional and Technical Writing program as well. The Online BGS allows students who have left the Georgia university system to return to finish their degrees online. Though the opportunity to develop online areas as part of the BGS was presented to the entire university, the only programs to respond so far are Business, Justice Studies, Sociology, and Writing. Currently, due to the program's overall success, other departments on campus are examining this opportunity. Beforehand, the professional and technical writing courses, other than the service courses, struggled to fill with enough students (15) to be taught; as part of the BGS, these courses now fill to capacity. For Professional and Technical Writing and the Online BGS, we currently offer these courses:

- WRIT 2130—Technical Communication—A service course developed principally for IT, electrical and electronic engineering, and construction management students.

- WRIT 3220—Foundations in Professional and Technical Writing—A common body of knowledge theory course required for all writing and linguistics majors. Minors and IT-second discipline students may also take it, though it is not currently available for BGS students.
- WRIT 3230—Writing in the Workplace— A service course developed principally for business education, hotel and restaurant management, and military science students.
- WRIT 3233—Professional Editing—Focuses on the job of an editor.
- WRIT 3030 & WRIT 5030—Selected Topics in Writing—Many courses such as "Writing Grants and Proposal," which will be in the 2013–14 catalog, began as "Selected Topics" courses. WRIT 5030 is for advanced undergraduates and graduate students.
- WRIT 3232—Document Creation for Writers—The title of this course was originally intended to be "Visual Rhetoric," but both the Communications Arts and Art departments opposed the title and elements of the proposed course description. The approved title was "Document Creation for Writers," which was changed this past year to "Writing Visual Information" for the 2013–14 catalog.
- WRIT 3234—Research Methods for Writers—focuses on ethnographic research.
- WRIT 4790—Internship in Writing and Linguistics—allows students to receive up to 6 hours of internship credit hours.
- WRIT 5550—Technologies of Writing—traces the history of written communication, defined broadly, from the perspective of the technologies used to facilitate and/or disseminate it.
- WRIT 5650—Writing for Publication—encourages writers from a variety of disciplines to learn how to publish their work.
- WRIT 5930—Technical Writing—capstone course that can focus on a variety of topics. A portfolio is a course option.

Our program has developed alongside increased computer use on campus. In the mid-1990s, prior to our department's existence, most universities created policies regarding faculty and student Web pages and email services. The university minimally encouraged Web page creation: students could create individual Web pages, and faculty could maintain course materials online. Since then, the university has faced policy decisions that have challenged faculty and student agency to create online materials. For example, when database-driven websites became feasible for those without extensive computer programming knowledge, the university decided not to support these open-source options, though it had supported and still supports other open-source options. Because the concerns were with security, database-driven options were not available to faculty and students who wanted to learn how to create more contemporary

Web pages. Faculty had two choices: limit their own sites and their pedagogy to basic HTML pages for websites or migrate website content to an off-site Web hosting company. As of 2012, only some of the university's websites were supported for content management systems, yet increasingly, websites at other institutions are database-driven. And as we well know, other institutions have faced similar problems.

WRITING PROGRAMS AND COMPUTER TECHNOLOGY

Issues of institutional security and individual agency have certainly been discussed in the technical communication literature. Stuart Selber (2009) has noted that, as a field, computers and composition specialists have "tended to focus more on students, teachers, classrooms, and programs and less on the larger contexts within which people and programs are embedded" (p. 13). His suggestions for how we may address "the interests and stakes of academic computing in ways that are meaningful to composition teachers" (p. 13) call for us to involve ourselves more in these important discussions, since, as he concludes, "In theory and oftentimes in practice, few other disciplines will know more about how the infrastructure might be employed for productive purposes" (p. 32).

Selber expands upon James E. Porter, Patricia Sullivan, Stuart Blythe, Jeffrey T. Grabill, and Libby Miles' work (2000), which, according to Selber (2009), offers a basis for "productive social action," because "institutions are not totalistic, unalterable entities governed by the unilateral actions of an elite few" (p. 13). Change, then, is possible "if people conceptualize institutions as a rhetorical system, pay attention to its contexts and constituent parts (including operating procedures and working conditions), and acknowledge their own involvements and commitments" (p. 13). Though they acknowledge the situated authority of these bodies, they remind us they "are not monoliths; they are rhetorically constructed human designs (whose power is reinforced by buildings, laws, traditions, and knowledge-making practices) and so are changeable. In other words, we made 'em, we can fix 'em" (Porter et al., 2000, p. 611). Ultimately, however, in the 21st century, power resides with those who wield the purse strings. Thus, when budgets are crunched, administrators can dictate technology expenses in ways that directly affect our pedagogical, curricular, and scholarly pursuits, often without consulting users until decisions are already made.

What we question are decisions being made solely by information technologists and administrators. While many of these decisions affected faculty, students, and pedagogical practices, it did not involve faculty—or students—as decision makers. Nonetheless, by paying attention, as Cynthia L. Selfe (1999) has reminded us, and becoming involved, as Selber (2009) and others prompt, we as stakeholders in this fray may at least make sure that the rhetorical and pedagogical underpinnings of technological decisions are heard, even if they are not always fully acknowledged.

These types of issues had certainly been anticipated. In 2007, Andy Guess predicted, "there is plenty of reason to believe that over the next few years, third-party Web services will become standard at most colleges and universities." He foresaw that this could have both benefits and drawbacks: though universities could relinquish some storage issues, software updates, and data security for faculty and students, "it does so reflecting Google's priorities and those of its many customers" (Guess, 2007), not necessarily those of universities. Universities, he continues, should carefully consider the privacy issue, especially in terms of the "consequences of moving students' and faculty members' personal data to off-campus servers over which they have no control" (Guess, 2007). Indeed, as Cecelia Kang notes, "Google will soon know far more about who you are and what you do on the Web" (2012).

Ironically, while many teachers and librarians steer students away from doing research exclusively with Google, many university-system IT administrators are pushing students deeper into the world of Google. While we readily admit to using Google and its products ourselves, we nonetheless decry the decision to foist it upon us—and upon our students—without first consulting us about the possible curricular ramifications.

A university's policies should account for faculty and student strategies for contemporary practices. As the university determines its support, faculty and students create attachments according to these policies. Those attachments then contribute to students' attitudes toward changing options for Internet access. For example, in the 1990s, websites became accessible for those who were not interested in learning extensive computer programming, but who nonetheless enjoyed learning how to create Web pages courtesy of our institution's hosting options. We might not have made these attempts if the university had not provided these resources. Hours of learning shape current attachments and relations to traditional websites founded on HTML coding and may extend from contemporary database-driven resources to those for mobile phones. Consequently, we harbor attachments to that literacy acquisition that contribute to expectations for continued university support.

Our curricular decisions are partially tied to the university's trending policies, which have included an increasing reliance on external services to manage technological needs. A series of contracts with corporations offering online platforms for a variety of services indicate the university's strategies for managing resources. Three factors seem to predominate.

First, some policy trends are determined by how funding shapes these decisions. Stipulations on the student technology fee, a large part of the annual budget for technology, control of how money can be spent, and external platforms fit with funding more easily than the university providing the resources.

Second, security problems have increased as more people place more private information online. These external platforms afford the university the comfort of a secure resource.

Third, the university constructs its role differently than in the 1990s. These financial decisions have resulted in increased class sizes and increased expectation that faculty will develop course materials for online delivery through a university-mandated course management system.

However, these newly implemented policies do not adequately address specific needs or digital literacies, especially for a meaningful technical communication program.

INTELLECTUAL PROPERTY

Intellectual property should certainly be part of the knowledge base of any budding technical writer. With that in mind, one of our faculty members has twice taught a selected-topics course on intellectual property that attracted majors in writing and linguistics, information technology, journalism, and others. The course provides an introduction and general overview of the fundamental aspects of intellectual property, and research and discussion of such related issues as music downloading, remix, plagiarism detection services, copy-left, open-source, and creative commons licensing.

In the past, students were asked to create a website to share their research. When students in this or other courses are taught to create websites, images are certainly important to discuss in terms of intellectual property: What constitutes "fair use?" How does one give credit for such "borrowings?" What are the possible legal and ethical ramifications of downloading images to use on one's website versus linking to images on someone else's site? If a link, for example, is created to an image on another website, then what does this imply about documentation and intellectual property? Whose responsibility is the image's documentation, the one who links to the image or the one who has posted the image? Students may also need to consider potential problems of inline linking (aka bandwidth theft) versus risks of downloading and republishing copyrighted images. If these are "academic" uses, is it okay? If we are teaching our students to be professionals, what will they need to know? The courts have ruled that the legal responsibility for Internet images lies with who posted the images, not who links to it, since the link is code, not an image. It is also worth pointing out to students that whoever posted the image might take it down or alter it. A group called Switcheroo offers this service. A good example of their action would be a heavy metal rock fan who was using a black fist outlined in red and set against a red background as an icon, one that had been appropriated from a heavy metal rock band; Switcheroo changed the inline linked file to a Brittney Spears poster. Inline linking can also increase operating costs for whoever posted the original image since the number of hits on the website can increase to the point that the owner of the site where the image is posted may be forced to increase bandwidth to accommodate the traffic.

With the Google suite, there are other, broader intellectual property issues that students in the Intellectual Property class had not been aware of—and didn't like when they conducted research into how the changes at Georgia Southern University affected their own intellectual property. Shifting toward corporately developed online platforms has been repeatedly supported, for example, with little regard to individual intellectual property rights as they relate to faculty and students, particularly the extent to which their data will become part of bigger data pools, including our university's use of WebCT for the past 12 years and a more recent shift to a new CMS that includes Turnitin. It has been debated in the courts as to what extent Turnitin can be allowed to profit from student essays that become part of their database, with faculty requiring students to provide the labor, at no cost to Turnitin, to upload to this database. Nonetheless, some faculty and administrators laud the use of such plagiarism detection services, often without considering negative pedagogical implications or the controversies surrounding use of students' writing for commercial purposes. For faculty who develop courses and provide course materials that are then made available to students in the CMS, it is also not always clear who owns the intellectual property rights. Certainly, a case can be made that the development of such courses is "work for hire" and therefore owned by the university, but under what terms could the CMS also claim rights?

Shifting to corporate services such as Google Sites for website creation requires universities to agree to the terms those corporations stipulate. For example, the CMS now in use at many institutions, "Desire to Learn," can interact with Google Apps for universities, potentially providing Turnitin with essays gleaned from our courses, along with sharing capabilities from a limited number of publishing companies such as Pearson (increasingly invested in creating lucrative testing/assessment resources). In addition, data can be shared much more easily between corporate platforms because the number of corporations offering them is shrinking, and there are considerable financial incentives for collecting that data. When a university agrees to trade data for security and cost reduction with a corporation like Google, faculty and students have little control over the current or future use of this data. For example, in January 2012, Google announced that it was changing its privacy settings to allow different applications to interact more freely and to allow Google to gain a "fuller portrait" of each person (Google Apps, n.d.). Now Google encourages users to link all their gmail accounts, both on and off campus. It is unclear how this information will be used, but it extends the implications of the university's agreement with Google.

While the university-system attorneys have focused on Google's privacy policies, and have agreed to them, what bothers us, but evidently does not bother our University Counsel or our technology gurus, are the rights of faculty and, even more importantly, those of our students. Google's "Terms of Service" stipulate,

> By using Google's products, software, services or web sites . . . [we] agree to the following terms and conditions, and any policies, guidelines or amendments thereto that may be presented to [us] from time to time, including but not limited to Program Policies and Legal Notices. (Google Apps, n.d.)

Our students have no choice but to agree, since this is now the "official" email provider for Georgia Southern University, where all official university correspondence will be directed to students.

Google's Rights, according to the document, include the following:

> Except as expressly authorized by Google or other proper third party rights holders, you [the user] agree not to modify, rent, lease, loan, sell, distribute or create derivative works based on Content, Google services or Software, in whole or in part except as specifically authorized in a separate written agreement. (Google Apps, n.d.)

However, the user's rights, as set forth in the next section of this contract, state,

> By submitting, posting or displaying Content on or through Google services which are intended to be available to the members of the public, *you grant Google a worldwide, non-exclusive, royalty-free license to reproduce, adapt, modify, publish and distribute such Content on Google services for the purpose of displaying, distributing and promoting Google services. Google reserves the right to syndicate Content submitted, posted or displayed by you on or through Google services and use that Content in connection with any service offered by Google* (Google Apps, n.d.; authors' emphasis).

Google's policy is far too similar to Turnitin.com's, which allows the company to exploit student work for profit. How has this been addressed in the contractual agreements negotiated by universities for use of Google's educational version? Further, like so many U.S. institutions, we are looking at expanding our university beyond U.S. borders via online course offerings and other means. What happens if our students are located in China, where Georgia Southern maintains partnership relations with other universities, or if other countries may block or limit access to Google? Will there be provisions made? What will they be? We believe it is incredibly important that we address these issues before we consider what we ask our students to do as part of a course, especially as this university, like so many others, begins to increase its offerings online. Not only is the professional and technical writing area heavily invested in Georgia Southern's Online Bachelor of General Studies program, but the administration is also interested in having our area develop an online graduate certificate in professional and technical writing, which will be marketed internationally. And these problems with issues of intellectual property feed directly into issues of privacy.

PRIVACY ISSUES

Privacy becomes an issue for students and faculty alike when their work becomes part of a database. While the Family Educational Rights and Privacy Act (FERPA) guidelines purport to protect certain student information, including but not limited to grades, transcripts, course schedules, and other information, the realities of the political climate affect our ability to make decisions about the best ways to protect such information, or even if we can still protect it at all. Moreover, when university resources shift, faculty must adapt to the realities of the climate at state universities braving economic downturns and state legislators increasingly attentive to individual professors because surveillance technologies have also improved. Amalgamating large amounts of data relying on sophisticated algorithms has already interrelated with increasing capabilities by government intelligence to analyze data. The strategies for surveillance that evolved in the post-9/11 framing affected attitudes toward security, toward banking and using credit cards online, and as people became much more willing to share personal economic information online, concerns about security became even more highly staked. While the decisions one university makes within the constraints of security, funding, and class enrollment shifts are clear, the best responses to these issues for those teaching digital literacy and writing for contemporary environments, are unclear.

In the midst of the local political climate, and the decisions that have happened regarding technological support, one option to address privacy might include the decision to migrate away from university resources for Web page design, which is one way to address the frameworks afforded by the university and to avoid how Web page content could be read by those who would like faculty to separate politics from work, as if politics can be ignored for rhetoricians. For example, in the 2011 case of Wisconsin vs. William Cronon (A Shabby Crusade, 2011), emails were part of an open-records act request. Cronon describes a strategy that faculty are increasingly adopting regardless of whether their work could be justified within university policies. He has chosen to maintain privately owned computers "because I haven't wanted to worry about whether my personal and professional emails are mingling on a state-owned machine"; but he adds, "The boundaries between these two categories is [*sic*] harder to draw for a scholar of the modern United States than non-scholars might imagine" (Cronon, 2011).

As little as 15 years ago, paying rent for space on a private server to help students create Web pages, maintaining a powerful Internet hook-up at home, or wondering whether to use any university resources and the consequences of each act in terms of the views of state legislators would have perplexed us. So what happened? There was a war, and the accompanying increased technological capabilities that facilitated surveillance included improved memory capabilities for a range of cameras and more sophisticated algorithms that could predict with amazing accuracy the purchasing moves for individuals. Though the ramifications are part of the macrocosm, the debate has failed to be cogently articulated for the microcosm. Obviously, our decisions have ramifications for our students as well as for ourselves. Is privacy a right for any of us in a digital age? What happens when we require students to create—and make public—websites that may deal with controversial topics?

Our university's answer seems to be that our students should not make their work public—keep it inside the course management software or, in Google Sites, share it only with those inside the university or for a given class (of course, this avoids the question of what rights Google might have to use the information). And yet many of us in technical communication saw the Web as a way of helping our students realize a greater audience than within a single classroom as part of our mission to help them become professionals.

PROFESSIONALIZING STUDENTS

We are, of course, concerned about the limits being imposed on our ability to professionalize students in technical writing. While Google Docs enables easy sharing of documents in a variety of media, not all applications and advanced features communicate seamlessly with Microsoft Office or other such proprietary products. While Google Docs certainly should be one option for students, we hope that universities will not consider this as a replacement for MS Office suites installed in our labs. Additionally, the opportunity to work with servers to create websites provides students with the ability to acquire valuable workplace skills, but this too is one of the issues that the adoption of Google Sites precludes.

Up until the shift to Google Apps, we had access to space on a server at Georgia Southern. Upon their enrollment, students also had access, but the space would not be created until the server received a command. This, we thought, was good practice for students who might someday need to set up or update a website for a client. They needed, we reasoned, to know how to interact with a server.

Working with websites is certainly one aspect of how we had been thinking about our students implementing as professionals what they have learned in our program. We wanted them to know how websites work, especially as this process relates to what should be their primary area of expertise, writing. For those of us who were teaching our students how to create websites, even in first-year writing classes, the opportunity provided them with a type of agency on the Internet and responded to a literacy issue. Digital natives, that is, tend to be users, not authors, of Internet content, at least in our experience.

In general, students responded favorably to such assignments. After setting up space on the server, they could be taught about HTML code with an assignment that required them to type the code in Notepad, or they could be taught to use one of the available Web authoring software packages such as Composer or Dreamweaver. Teaching HTML allows for a great deal of conversation about Web pages and the decisions that must be made to create one. For example, using <body bgcolor="#FFA07A"text="# 000000"> will render a “light salmon” background, according to HTMLCodes.ws (2012). Some students immediately objected to that color, allowing for a conversation about color in general, color-blindness, and the problems this color could create for the color-blind, about 13% of all men and 8% of all women. The code <img src= “http://www.georgiasouthern.edu/~tgiles/ Beach.gif”> would place an image on the page, which could lead to discussion about the lack of documentation for the image, how it might be inserted, and the correct disciplinary approach for doing so.

The image itself was worth discussing. Students could be asked how the image could appear there. Usually some students could point to the HTML code that indicates the image appears on another website. However, going to this website would not reveal this image. The professor could then explain that the file is stored on this website, but not currently being used on it.

Other issues concern the Web page format students have created through the exercise. For example, in the HTML exercise, they give their Web page a file name and a title, which would be centered because of an HTML command. How commands work in terms of opening and closing tags could be discussed as well. Such an exercise could also be used to create links to other pages, bulleted lists, email links, and others.

An email address code allows for rich discussion with students as well. Though some programs such as Dreamweaver contain a tool that will create an email address link, Sea Monkey’s Composer requires that users understand how links work (using the mailto: instead of the http:// protocol, for example). This fact in itself is a point for discussion: first, Dreamweaver costs hundreds of dollars, but Composer is free; second, knowing the code gives novice Web page creators greater agency in terms of what they can do and how they can do it. Even though it is possible to edit code directly in both Composer and Dreamweaver, IT students at Georgia Southern who were once taking courses in Dreamweaver currently are exclusively learning coding. When IT students were learning Dreamweaver, they would find that some tasks, such as creating columns, were easier to do by coding rather than figuring out the complexities and idiosyncrasies of the Web authoring software.

After students developed some familiarity with HTML, they could be taught to use Sea Monkey’s Composer as a Web authoring tool, moving through the same process in a fraction of the time, of course. It could also be instructive to show them how much more garbage code may be included with a Web authoring tool, versus Notepad, for example, as well as how much more accumulates when an HTML page is created with Microsoft Word.

Typically, students would want to include images on their Web pages, which is instructive, again, for review of intellectual property issues and documentation, and allows for explaining the use of Creative Commons images versus those located through Google. Students could also learn to choose images according to size since a bitmap image could easily exceed the amount of space they were allotted on the server and clog up their email, which reinforces the notion of interacting with a server. Layout and design could be approached; students could be taught how to use tables or, better, CSS to keep images from sliding around and to keep their Web pages from looking too radically different from fallout from the Web browser wars. And of course we want our students to understand issues of accessibility (e.g., the use of "ALT" tags, how screen readers work, how colors may appear to the color-blind, etc.).

Such a classroom activity can be useful for first-year composition students who have not decided what they want to major in, and of course we want to graduate technical communication students who have these skills. We had been introducing our other writing and linguistics students who are not primarily interested in professional and technical writing to these ideas as well; a creative writing student might very well be interested in starting an online poetry journal.

Web page skills might have been expected to take the same turn as word processing in terms of transparency. For example, when students had access to computer lab classrooms in the 1980s, they typically had to be taught word processing skills. By the early 1990s, asking a class who knew how to use a word processor would yield a show of a few hands. As the decade progressed, the question about word processing skills no longer needed to be asked, though it would sometimes be apparent that the occasional student did not know how to do word processing, or that there were smatterings of students from the same town who thought the way to double space a document was to hit the return key between each line. Indeed, at the turn of the millennium, some students were learning how to create Web pages, and more about computers, in high school classes, but that trend seems to have waned. To some degree, ironically, the popularity of social media sites might be blamed, along with the IT developers' ethical mandate to always remember the human and to create evermore intuitive interfaces. Those facets, coupled with security and the desire to save money on support, have created an atmosphere wherein we have lost a server. We have gained Google Sites, though.

There are, of course, problems with Google Sites. It makes decisions for us, such as stripping out the salmon background color when that page is loaded. The linking tool is not intuitive for someone who is used to doing any type of linking. An animation that worked on the desktop and has easily worked online did not work on Google Sites. It is still possible to manipulate the HTML code with a page loaded onto Google sites, but as a Web authoring tool, it is somewhat cumbersome, and the interface is not intuitive. Of course, it is possible that our students will need to be able to work within the limitations of such templates when they enter the workforce. However, the more important question is, do we want to graduate technical communication students who know how to create a website with only Google sites?

POSSIBLE SOLUTIONS

Migrating Web page content off the university's servers allows students to be introduced to alternate ways of creating Web pages, but its value depends upon how the migration takes place and where the user winds up. While students continue to be taught how to create HTML-based Web pages, work can also take place in content management spaces to meet other curriculum outcomes. We wanted students' experiences to be similar to what they would encounter when they graduated from the university.

In 2005, one author argued that it would make the most sense for our university to create the equivalent of Web hosting sites available on the Web so that we could teach our students how to use these resources in the future. If we created a sort of look-alike Web hosting "company," students could learn how to create basic Web pages, choose a Web address, and decide if they wanted to try out various open-source options. They might create, with one-click installation, a Wordpress blog, or an e-commerce interface, and then they could learn to create a viable, working online space. While this solution would not be ideal, it could have been one response to the shifting realities of our university's decisions for Web page support. Unfortunately, the resources were not made available.

While we might hope for our institution to fulfill a relevant curriculum, austerity prevents it from doing so fully, so the question becomes, how do we make sense of attachments to abstract and concrete entities that shape how we respond as educators? These issues may result in a tendency to explore attachment to options no longer available, in part because we've outgrown them, but in part because the university has faced the hazards of funding pressures and has become much more defensive, much less able to provide for its faculty and students because it has had to rely on other ways of funding and therefore has to be cautious in how it defends its resources. Ultimately, its decisions regarding technology, security, and access are part of the economic downturn resulting from intertwining wars and the banking crisis. Ironically, our curriculum collides with the challenges to the technology intended to support people.

Ideas or paradigm attachments stack up and become antiquated. How we as writing teachers value theory parallels Thomas Kuhn's (1970) work on shifting scientific paradigms that herald revolutions of belief, or more recent explorations about how our memory and our beliefs intersect, how our ways of thinking, or beliefs, interfere with new information that contradicts what we think we know. Kuhn, whose PhD was in physics, was thinking about the paradigm shifts that occur in science. Composition, too, has had its paradigm shifts, beginning in the 1970s when writing began to be taught as a process rather than focusing on the product. In the 1980s, as computers became part of the environment, the writing instructor became a teacher of hardware and increasingly, of software, a trend that only accelerated in the 1990s. Today, early in the second decade of the new millennium, yet another paradigm results from the IT professionals' mandate: Remember the human, in the sense that IT development leans toward the intuitive, and what is intuitive transforms as the cell phone and the e-tablet merge. This leaves us to wonder about the point of teaching, and learning, new software program after new software program, when the time, and more importantly, the resources, simply either are not there or are limited in some fashion. To what extent should we remember that we are writing professors—those who profess writing as a distinct academic area of inquiry—not IT professors? How can we balance these concerns with offering students what they need to succeed in a world whose technological evolution has accelerated, and will only continue to do so?

Deborah Brandt's (1995) accumulating literacies are useful as a means of articulating and arguing for various literacy positions. How might we clear out mental space and do away with the foundational thinking that settles into accepting Brandt's accumulated literacies? How do we clear a path, migrate more easily from one way of thinking to another? How do we relinquish expectations for how the university should support technology? Do we insist that the university support and fund work in technology that ties into where expectations for academic freedom are much clearer? What is the need to accumulate literacies, and how do we think about strategically clearing mental space, detaching from ideas and ways of being that no longer hold? Do we need a way to do this that remains aware of nation-state and post-nation-state corporate framings that shape what we value and what we believe to be important ways of thinking? It matters, at this point, to move back and forth between the approaches we have toward books and technology and to argue for similar ideals regarding the opportunities for the thinking and experimenting that the university supports, but without insisting on a trade of support for data collection. The best compromise may be to argue for an external platform for our students—a Web hosting site that allows students to learn these resources, but on the university's dime, not ours.

CONCLUSION

In this age of austerity, our university is pushing faculty and students into Google Docs, Google Presentations, Google Sites, and even Google+. Of course, while introducing some Google products to students is certainly worthwhile, nonetheless, at this juncture, there are also extensive limitations beyond those discussed in this essay. However, we have no choice if we are required to use Google Sites for our Web pages and Google-compliant artifacts in this new era of the Google University.

As faculty, we believe an important part of our job is to help students learn to make informed choices. To that end, we want to prepare students for the workplaces they are likely to find themselves in, but we also want to prepare them to make decisions about those workplaces as the available tools continue to morph. And of course we would like to introduce our students to open-source and open-access tools as well—not only those tools that depend upon and help support corporate sponsors. In the case of our university, the decision to use Google Apps is fraught with unanswered questions about privacy, intellectual property, and of course teaching and learning.

Some of our students have investigated these resources and are using them. They even seem to like these tools. But few have read the terms that they agreed to simply by using them. If they have, they do not like the fact that, even though they do own the copyright to their work, they are "licensing" Google to use this work for marketing purposes. They want to control what use is made of their artifacts. However, they have no choice: the university agreed to these terms for them.

As Jonathan Zittrain (2008) notes in *The Future of the Internet and How to Stop It,* we may be swiftly moving toward a

> technological ecosystem in which non-generative devices and services—sterile "tethered appliances"—come to dominate. This trend threatens to curtail future innovation and to facilitate invasive forms of surveillance and control. A non-generative information ecosystem advances the regulability of the Internet to a stage that goes beyond addressing discrete regulatory problems, instead allowing regulators to alter basic freedoms that previously needed no theoretical or practical defense. (p. 63)

Georgia Southern University's move to "enhance" security by outsourcing so much of the core services such as Web page hosting is a move toward tethering not only our appliances but our entire institutional—and perhaps our private—lives.

We are moving ever closer to a world where nothing the individual does or says will be private in any sense of the word, and we may very well be limited to using only what we are allowed to use. Cloud computing, for example, while useful in many ways, can also be used to control users. Will we again see the days of "dumb terminals" instead of full-featured computers, so instead of accessing programs on a mini-computer, we will access them only online? Many universities and companies already do this, saving money on software licensing fees and IT expenses, while making more programs available to more users.

We might also question, though, to what extent such dependency is realistic. On the subject of social media sites, Eric Jackson (2012) has raised a more important point in his article whose title speaks for itself: "Here's Why Google and Facebook Might Completely Disappear in the Next 5 Years." Jackson begins by explaining organizational ecology, a theoretical approach that posits how a business is managed is not as important as the environment in which it functions. A concept pioneered by sociologists in the 1970s, organizational ecology accounts for the "birth and death rates of populations of organizations, as well as the effects of age, competition and resources in the surrounding environment on an organization's birth and death rate." Considering the history of the Internet and e-commerce, Jackson defines Web 1.0 as characterized by the greater diversity of Yahoo, AOL, Amazon, and Google. With these four, we saw the emergence of social media in the now quaintly antiquated AOL, e-commerce in the shape of Amazon, Yahoo's attempted merging of social media with a better search engine, and the extent to which the public indeed valued a better, sleeker search engine (with the news, weather, and sports available, if and only if, we wanted them) in the shape of Google.

Jackson (2012) narrows Web 2.0 as defined by Facebook, LinkedIn, and Groupon, which limits Web 2.0 to social media and e-commerce. And Web 3.0 poses the question of survival for Facebook and Google: To what extent is either relevant as an app? Each has made some attempt to be one, but doing so fragments the original offering; or, as Jackson notes, to some extent their original business plan defeats them. For technical communication, then, teaching students to use Google Sites today might put them at something of a disadvantage if Google Sites becomes the equivalent of telephone booth stuffing or goldfish swallowing.

If, however, *all* software is accessed online, and all of our files are stored there, then not only might it be possible to track all use, but we will no longer be able to "borrow" software from friends; we may no longer have access to create our own software packages; and we will no longer be able to decide for ourselves to keep using old, unsupported versions of software, but will be forced to upgrade when the online provider decides. In "The Coming War on General Computation," Cory Doctorow (2011) argues, "The coming century will be dominated by war against the general purpose computer, and the stakes are the freedom, fortune and privacy of the entire human race." Such a war does not need to be fought if the issues can be successfully negotiated, and while we have such a framework for negotiation written into our political life, it is fraught with influence, especially from corporate sources. It cannot be within the scope of one essay to decide how to address these issues, which, as we have noted, are always contextualized by individual circumstances. We can, however, continue to search for new arguments to support our struggle for our rights, and for the rights of our students.

REFERENCES

A shabby crusade in Wisconsin. (2011, March 25). Editorial. *New York Times*. Retrieved from http://www.nytimes.com/2011/03/28/opinion/28mon3.html?_r=0

Brandt, D. (1995). Accumulating literacy: Writing and learning to write in the twentieth century. *College English, 57*(6), 649–668.

Cronon, W. (2011, March 24). A tactic I hope Republicans will rethink: Using the Open Records law to intimidate critics. *Scholar as Citizen*. Retrieved from http://scholarcitizen.williamcronon.net/2011/03/24/open-records-attack-on-academic-freedom/

Doctorow, C. (2011, December 27). The coming war on general computation: The copyright war was just the beginning [video]. *28th Chaos Communication Congress.* Retrieved from https://www.youtube.com/watch?v=HUEvRyemKSg

Georgia Southern University. (2009, June 16). Top ten things to do with MyApps [weblog]. *MYAPPS at georgia southern.* Retrieved from http://georgiasouthern googleapps.wordpress.com/2009/06/16/top-ten-things-to-do-with-myapps/

Georgia Southern University. (n.d.). What is Google Apps? *MYAPPS at Georgia Southern.* Retrieved from https://my.georgiasouthern.edu/portal/myapps/

Google Apps. (n.d.). *Terms of service.* Retrieved from http://www.google.com/apps/intl/en/terms/user_terms.html

Guess, A. (2007, November 27). When e-mail is outsourced. *Inside Higher Ed.* Retrieved from http://www.insidehighered.com/news/2007/11/27/email#sthash.fO6aWmkv.dpbs

HTMLCodes.ws. (2012). *HTML codes.* Retrieved from http://www.htmlcodes.ws/

Jackson, E. (2012, April 30). Here's why Google and Facebook might completely disappear in the next 5 years. *Forbes.* Retrieved from http://www.forbes.com/sites/ericjackson/2012/04/30/heres-why-google-and-facebook-might-completely-disappear-in-the-/

Kang, C. (2012, January 24). Google tracks consumers' online activities across products, and users can't opt out. *The Washington Post.* Retrieved from http://www.washington post.com/business/economy/google-tracks-consumers-across-products-users-cant-opt-out/2012/01/24/gIQArgJHOQ_story.html

Kuhn, T. S. (1970). *The structure of scientific revolutions* (2nd ed.). Chicago: University of Chicago Press.

Porter, J. E., Sullivan, P., Blythe, S., Grabill, J. T., & Miles, L. (2000). Institutional critique: A rhetorical methodology for change. *College Composition and Communication, 51*(4), 610–642.

Selber, S. A. (2009). Institutional dimensions of academic computing. *College Composition and Communication, 61*(1), 10–34.

Selfe, C. L. (1999). *Technology and literacy in the twenty-first century: The importance of paying attention.* Carbondale, IL: Southern Illinois University Press.

Zittrain, J. (2008). *The future of the Internet and how to stop it.* New Haven, CT: Yale University Press.

CHAPTER 10

Strategic Partnerships Promote High-Demand Technical Communication Courses

Lynn O. Ludwig

Many colleges are redefining and expanding their mission and goals from the top down to ensure survival in economically challenging times. While the direction may come from a hierarchical infrastructure, collaborative implementation of the change is necessary. For truly successful redirection, the transformation requires all vested parties to participate actively in growth and expansion within each step of the process. The focus of this chapter is on the significant changes currently underway within the University of Wisconsin (UW) System, specifically highlighting one particular campus: the University of Wisconsin at Stevens Point (UW-SP). State-mandated educational goals are directing the programmatic and pedagogical changes within the UW System offerings, and the UW-SP campus has embraced the opportunity to model a collaborative spirit of change. This chapter will apply a framework of growth and change to developments currently underway at UW-SP to create a new course aligning with a broader community health education initiative. The growth is described in three phases: awakening, pioneering, and transformation.

The town of Stevens Point (population ~27,000), known simply as "Point" to locals, is considered "The Gateway to the Pineries" and boasts spectacular views of majestic pines within and around the town's boundaries. The beautiful Wisconsin River runs through the town and provides numerous recreational and leisure activities for residents. UW-SP students also enjoy Schmeeckle Reserve, a

280-acre natural area and College of Natural Resources field station on the campus that "provides rich learning and research opportunities for all UW-SP faculty and students" (UW-SP, 2010). UW-SP programs collaborate with the residents of Point toward ongoing efforts with conservation and sustainability. The UW-SP campus is strategically located in the heart of Wisconsin, which allows for convenient working relations with other UW System campus communities. The school attracts applicants from around the state and beyond and offers associate and baccalaureate degrees, as well as select graduate programs.

An important focus has been acknowledged and embraced in Wisconsin—a focus on today's student profile: the underrepresented and nontraditional. In Morey's (2004) exploration of globalization and the for-profit educational model, she explains, "Higher education has gone from an elite, mainly private system to an open, public system of mass education. One important shift has been the entrance of non-traditional students, including adult students and those from underrepresented groups" (p. 147). Higher education is offered in multiple locations throughout the state of Wisconsin, with enrollment encouraged toward nontraditional and underrepresented populations. The UW System offers quality education throughout the state, including 26 campuses (UW System, 2012). The expansive UW System schools are under the direction of the governor of the state, and as the governor directs, the System responds. UW-SP has readily embraced efforts to redirect itself to align with the needs of today's student and tomorrow's workforce, including concentrations in various healthcare initiatives. Current and future plans for new technical writing courses are an important part of those efforts.

PROFESSIONAL WRITING AT UW-SP

The UW-SP Department of English offers numerous courses for English majors and minors. The professional writing minor (PWM) requires students to take 24 credits beyond the general education composition requirement. The PWM is structured around 3-credit courses that include, but are not limited to, the following: Introduction to Technical Writing, Advanced Technical and Scientific Communication, Advanced Business Writing, Editing and Publishing, Grant and Proposal Writing, Freelance Writing, and Biomedical Writing. Upon completion of the PWM, students will

1. demonstrate in their work an awareness of the power of language to influence readers' thoughts and behaviors;
2. employ appropriate writing strategies to respond to the needs of readers in various contexts; and
3. write creatively and effectively in one or more category of professional writing (business, scientific, technical, or biomedical) (UW-SP, 2012a).

For years, the department has offered a biomedical writing course, and a new biomedical writing minor is now available. The minor requires 23–24 credits, plus a 1–3 credit recommended internship. The new minor provides students a competitive edge in scientific and biomedical writing skills before they enter medical professions. Writing internships are also available and provide hands-on learning in the following areas:

- various opportunities in central Wisconsin, at city newspapers and publishing houses (for English majors),
- internships in London with public relations firms, publishing houses, and professional organizations (for English majors or writing minors),
- editorial internships for Issues in Writing, a nationally recognized scholarly journal published by the English Department, which features articles on writing in Education, Business and Industry, Science and Technology, Government, and the Arts and Humanities; and,
- biomedical writing internships available with the Office of Scientific Writing and Publication at the Marshfield Clinic Research Foundation (4–6 students/year). (UW-SP, 2013)

The current PW minor and internship opportunities provide solid ground for future PW courses and program offerings by the English Department. Through internships, students are introduced to the work environments they will encounter before they begin their careers in the different settings. To assist students in gaining entrance to medical settings, some courses will incorporate experiential learning (XL) activities to provide skills training that simultaneously benefit the student and the organization. Plans are currently underway for a new PW course that includes an XL component that will enable students to immerse themselves in the work environments they will encounter in their careers. The new course will possibly be an elective to support the existing PW minor and the newly created biomedical writing minor. Before exploring the new PW course in more detail, a framework is presented to situate the ongoing change-management process at UW-SP.

METHODOLOGY

This chapter applies a framework for growth and change to document what is currently underway at UW-SP, specifically geared toward the development of a new PW course designed to align with the larger statewide community health education initiative. Throughout this chapter, the "Organization Change Management Framework for Sustainability" is borrowed from Yale University and applied (Newman, 2012). This framework focuses on the following three phases: awakening, pioneering, and transformation, with a transitioning phase occurring between each of the three main phases (Newman, 2012). The framework provides guidance for vested stakeholders as they move through the arduous task of creating something new. The 3-step process accurately describes past and current steps taken by the larger UW system as it moves toward today's academic goals.

Newman (2012) aptly explains that growth opportunities are expected and planned for in this process as it "prepares the institution for course correction and the ability to continuously seek new solutions and set new goals over time" (Abstract). This approach allows for the following: planning the new path, reviewing successes and challenges from the past, modifying steps taken to help ensure successful growth toward goals, and periodically revisiting the entire process and its outcomes, with an eye toward revision, or as Newman says "course correction" (p. 67). More specifically, the changes underway with the UW-SP Healthy Communities Initiative will benefit from the application of this 3-part process. The framework includes the following phases:

1. Awakening: In this phase, the general public, as well as key stakeholders (those with a vested interest in the changes) become aware of and join in the activities toward change (Newman, 2012, p. 67). Their activities include, but are not limited to, public announcements through organizational and regional news agencies, faculty and administrator collaborations, and committee formation.
2. Pioneering: The phase in which the vision, direction, and problem definition are advanced (Newman, 2012, pp. 67–68). Innovation and accountability are critical elements of the decision process (Newman, 2012, p. 68).
3. Transforming: The phase in which new processes and structures are integrated into the existing model (Newman, 2012, p. 68).
4. Transitioning: The phase that occurs in between the three defined phases, for example, wherein the pilot course content and curricula move into the permanent structure, as maintained and owned by numerous stakeholders (Newman, 2012, p. 68).

The framework reveals the complexities of major organizational change, as well as the intricate details required for new development to occur. Its application is well suited for presenting current and future changes occurring in the UW-SP system-wide initiative. While the process may appear to be straightforward and uncomplicated, in practice, its shape and demeanor reflect the unique qualities of each institution seeking growth. This growth may occur much faster for some than others, but in all settings, dedicated individuals are mandatory for the process to ultimately succeed.

APPLICATION

Phase 1: Awakening

> It's exhilarating to be alive in a time of awakening consciousness; it can also be confusing, disorienting, and painful.
>
> —Adrienne Rich (n.d.)

In the Awakening phase, the general public, as well as key stakeholders become aware of and join in the activities toward change. Communication by institutional leaders is necessary to be successful in moving an entity toward new goals. UW-SP administrators have addressed the campus and local community numerous times. The need for change was presented in the annual State of Academic Affairs Address in 2012, stating,

> Public universities across the state must learn to adapt to the new business model for higher education or their very survival could be at stake . . . the confluence of events that now confront public universities, like UW-SP, render us one of the most vulnerable institutions in higher education. (Carlson, 2011)

The challenging task of redefining an institution's direction is often delegated to a number of stakeholder groups to create opportunities for collaboration and direction. UW-SP has made sure that public information is published online and in print, to help ensure clarity and visibility to all community members. The Office of Academic Affairs defines four specific special initiatives that are underway at UW-SP, including A Partnership for Thriving Communities. This initiative "provides the framework through which the university will become more engaged with area stakeholders, more responsive to local needs, and more relevant to solving regional problems" (UW-SP, n.d.). The goals of the initiative are to

- create a variety of academic pathways for students in the region—especially first-generation, underrepresented minority, and adult populations—to enter health-related fields,
- ensure students receive the best possible education by enhancing healthcare curricula and providing focused student support including intensive tutoring and advising, career counseling, and workshops and camps to assist with course work and professional exams; and,
- assist graduates in returning to central and northern Wisconsin to practice as health care and wellness professionals. (UW-SP, n.d.)

These goals serve as high-level guideposts for revising selective program and course content. Furthermore, the subinitiative, the Healthy Communities Initiative, "is a public-private partnership, intended to capitalize on the existing strengths and array of UW-Stevens Point's professional health care and wellness programs" (UW-SP, 2012b). The most significant goal of this subinitiative pertinent to this chapter is to "assure the best health care education possible by expanding health care curricula, workshops, summer camps, tutoring, counseling, and practicum experiences" (UW-SP, 2012b). Critical stakeholders in the health education initiative include current and future healthcare students, faculty and administrators in the academic settings, and administrators and organizational heads in the health facilities. All of these stakeholders play important roles in validating the appropriate content for the new PW course, and their input will be critical to guide the course as it evolves over time.

In-Between Phase: Transitioning

> Not in his goals but in his transitions is man great.
> —Ralph Waldo Emerson (n.d.)

The transitioning phase occurs in between the three defined phases (Newman, 2012, p. 68). By redefining higher-level goals, changes begin throughout the institution. The UW-SP Healthy Communities Initiative provides the motivation for an institutional transformation that creates a unique opportunity for individual departments to position themselves as leaders in today's liberal arts education. Anthony DiRenzo (2010) asserts that "if the humanities are to remain viable, dynamic, and relevant . . . liberal arts colleges must redefine their mission" (p. 244). Particularly appropriate to the current educational changes facing UW-SP is the idea that liberal education is a practical education. Some content may be shifting, when required, to meet the needs of a niche professional market, but overall, the same skills and knowledge are valued just as always. The courses that are transitioning to meet the new goals of the Healthy Communities Initiative exemplify DiRenzo's understanding of liberal arts by providing a practical education—a liberal education that "develops just those capacities needed by every thinking adult: analytical skills, effective communication, practical intelligence, ethical judgment, and social responsibility" (p. 245). The excitement heightens during the transition phase of any new venture when the stakeholders, that is, faculty members and program directors, visualize how the newly designed parts fit together to create the new whole. The culminating impact of program and curriculum changes (that meet the new needs of community members) will awaken all involved to the accomplishments of the partners that pioneered the transformation. As DiRenzo explains,

> It will have the strongest impact when studies reach beyond the classroom to the larger community, asking students to apply their developing analytical skills and ethical judgment to concrete problems in the world around them, and to connect theory with insights gained from practice. (pp. 244–245)

It is within this context that initial steps are taking place to align new curriculum goals with the needs of medical communities in north-central Wisconsin.

The potential for a new PW course is an exciting contribution to the larger state-supported program demands. Research into existing courses across the UW system identifies possible gaps that create new opportunities for the proposed course offering. Determination of course activities will take into account the learning outcomes of the existing UW-SP English Department PW curriculum goals, as well as the learning outcomes required by new academic goals. In the transitioning phase, plans for a new course are determined, and its placement is fixed within the existing curriculum structure. Collaborations with stakeholders in the department and the college are ongoing throughout this phase, with the goal of offering a new PW course.

Phase 2: Pioneering

> If we prove capable of showing a pioneering commitment, we shall create a community listened to around the world.
>
> —Jacques Chirac (n.d.)

In the Pioneering phase, the vision, direction, and problem definition are advanced (Newman, 2012, pp. 67–68). Innovation and accountability are critical elements of the decision process (Newman, 2012, p. 68). In the UW-SP setting, it is the point at which key stakeholders work toward the common goal of providing the learning contexts required by the new initiative's needs. In February 2012, listening sessions were held with numerous stakeholders, including education partners, healthcare executives, elected officials, economic development executives, and university faculty and staff (UW-SP, 2012b). A carefully selected team was convened that included a consulting group and a UW-SP dean emerita to prepare a study of the initiative to "serve as the blueprint for moving forward" (UW-SP, 2012b).

Just 4 short months later, in June 2012, the Medical College of Wisconsin (MCW), located in Milwaukee, announced that it "will offer new programs in satellite campuses, including Green Bay and Central Wisconsin, to address the projected shortfall of physicians and the maldistribution of health care providers in underserved communities across the state" (UW-SP, 2012c). After months of investigation, MCW announced their central Wisconsin selection: Wausau. Wausau is located just 30 miles north of Stevens Point and has a population of roughly 40,000, who benefit from numerous medical resource facilities in

Wausau and its neighboring communities. Those facilities serve the residents of the entire state and beyond. Due to its close proximity to the UW-SP campus, a partnership is readily available.

Furthermore, two additional healthcare alliances are currently underway. One involves UW-SP and UW-Marshfield, to develop a new Bachelor of Applied Arts and Sciences degree that focuses on health and wellness. The other unites the Mid-State Technical College (with campuses in Stevens Point and nearby Marshfield and Wisconsin Rapids) to UW-SP for the creation of a 4-year nursing degree to help RNs achieve baccalaureate-level skills.

Most certainly, partnerships are being forged between and among science and medical programs throughout the state to support the success of the new educational opportunities in health care. Specifically, PW curricula and course plans at UW-SP are being embraced with input from numerous stakeholders. The process of building relationships with key players requires time and effort. The first step is to find common ground that will result in meeting the goals of the initiative. Next, active stakeholders are identified through the media (local news broadcasts and online releases) and by talking to academic and medical administrators. Finally, contact with the targeted stakeholders initiates the spirit of cooperation and collaboration and provides the foundation of successful change. To plan and create the necessary changes in a select UW-SP PW course, relationships are being established with the following stakeholders:

- Science faculty at UW-SP (who have been identified by MCW).
- UW-Marshfield science and medical faculty (supporting the BAAS degree).
- UW-SP Nursing Program Director (supporting the BS in nursing degree).
- MSTC Associate Dean for the Service and Health Division (supporting the BS in nursing degree).
- MCW representatives (presenting and meeting with interested faculty from central Wisconsin campuses).

Ideally, an introductory meeting that includes all of the key players will occur in order to ensure commitment to the collaborative endeavor. In this meeting, each stakeholder should contribute to the specific and long-term goals as well as the necessary revisions and redirections throughout the process. This will help ensure the successful end results, that is, appropriate and accurate PW content and course activities. Key to the entire process is the inclusion of the students enrolled in the new programs and courses. Once the programs are in place and students have graduated and become gainfully employed, those graduates continue to be valuable stakeholders in the ongoing revision and redirection of the courses that support the knowledge and activities they perform on the job. This information-gathering is critical for the necessary revisions of the PW course, so the content and course activities remain applicable and sustain relevance. Students enrolled in the first offering of the new PW course will be asked to participate in follow-up interviews once they have established themselves in jobs that require them to use the knowledge and skills taught in the course.

RESULTS

As with all change, there is tension and fear. While reactions to MCW's campus selection were being discussed in central Wisconsin, the state's governor gave a speech in California on upcoming new policy initiatives that include "requiring Wisconsin's schools, technical colleges, and universities to meet certain benchmarks to earn state funding" (Hall & Derby, 2012). He proposed to "tie funding for technical colleges and the UW System with how well those institutions prepare students to take available and needed jobs in Wisconsin" (Hall & Derby, 2012). It is certain that individuals will disagree with the governor's education initiative, as one Democratic lawmaker stated after the announcement, the proposal "sounds like 'social engineering' that would force students to study 'what industry wants' rather than what students want" (Hall & Derby, 2012). This tension has long been the fuel for debate in goal discussions of higher education.

The Wisconsin governor's mandate, to tie higher education performance to the jobs that students take after graduation, appears to provide opportunities for some and the feared demise of others. It is important to consider that applied studies do not preclude the need for those subjects (and content) that reinforce the skill base of any given professional. Analyzing problems and articulating solutions to those problems are skills that can be achieved in all forms of study, while investigating multiple viewpoints and learning how to look at the world through the eyes of others can be achieved in the study of various subjects as well. The application of skills in a specific job niche is the next step in preparing the student for the professional settings they will encounter after graduation. Henze, Sharer, and Tovey (2010) shed light on the placement of applied skills in a liberal arts education by stating,

> What's up for grabs in boundary conflicts is not just traditional "resources" (such as faculty lines, research funds, courses, and students), but also control over representations of the discipline's central problems, concepts, and methods—that is, the "rhetorical resources" that disciplines create and maintain in order to solidify their boundaries. (p. 70)

Boundaries are necessary for institutional departments, but the existence of such boundaries does not prevent collaborations among relevant partners, most importantly, faculty members, administrators, and established employees, as well as others. Furthermore, Hyland and Kranzow (2012) argue that "the U.S. Department of Education (2006) recommends that colleges and universities embrace a culture of continuous innovation and quality improvement by developing new pedagogies, curricula, and technologies to improve learning" (p. 3). The cornerstone of success with UW-SP's process has been, and will continue to be, alliances and partnerships with many stakeholders, both inside and outside the academy.

While the programmatic changes are still in the early phases at UW-SP, monumental progress has already been made toward a "shift of consciousness" in understanding and embracing where the institution is and where it needs to be to continue to survive and flourish. At the same time, it is important to address some of the more difficult issues that are inherent in a changing environment. Successful pioneers of change understand that part of their role is to inform, educate, and enlist supporters and collaborators throughout the process. Initial contacts and outreach to colleagues may result in their hesitation and resistance of the discussions, especially if they feel threatened about the work it will take to make appropriate changes to their long-standing course plans. Furthermore, they may be concerned about the impact on them if they do not make the changes. These colleagues might not voice these feelings and choose to stay unattached to the entire process. Others might simply ignore or dismiss attempts to connect with them about the changes ahead. These are normal, expected responses to growth and change. Depending on the individual, attempts can be made to engage them personally rather than via email or phone calls. Introducing colleagues to time-saving technologies and tools that can be used to facilitate the changes may also help ease them into active change-agent roles. Enlisting their expertise on the impact of the newly revised courses on the existing curriculum often bridges them into an active role. Most often, illustrating how they will benefit personally in the advancement of new course goals will help ease their transformation into pioneers of the new program.

In the spirit of increasing opportunities for nontraditional students, the systemwide initiative was promoted to better serve adult learners in Wisconsin and a new Bachelor of Applied Arts and Science (BAAS) degree that focuses on community health and wellness was approved. UW-SP has entered a partnership with UW-Marshfield wherein their students will be allowed to complete courses from UW-SP toward the BAAS degree. Initially, there would be no significant impact to the teaching of the UW-SP courses, but eventually, to best serve the students and the new degree program, collaborations will be necessary with the UW-Marshfield faculty to align unit and program-level goals and learning outcomes.

Additionally, since MCW chose Wausau as one of "two community-based medical education campuses," UW-SP has begun to partner with UW-Wausau faculty to support MCW's mission (UW-SP, 2012c). MCW is "launching the community-based medical education initiative to address the shortage of physicians and other health care providers in Wisconsin, especially in underserved rural and urban areas" (UW-SP, 2012c). This announcement started a cascade effect that has already impacted course planning and curriculum design across many disciplines. UW-SP has articulated the upcoming high-level curricular changes:

> Plans call for the curriculum to teach "Triple Aim" core competencies: improving the patient experience (including quality and satisfaction); improving the health of populations; and reducing the per capita cost of health care. The curricular development also will include a focus on opportunities for interprofessional learning with other health sciences programs such as physician assistants, pharmacy, nursing, or dentistry to emphasize a team-based model of care, and leverage distance learning techniques. (Summers, 2012)

Numerous opportunities will come from the new direction, and each opportunity begins with small steps, taken by individuals, one course at a time. Through collaboration with vested stakeholders, specific curricular changes and additions involving individual course content must align closely with the needs of the students, the different programs they support, and the larger focus of the institution. These curricular changes will succeed when they support the emerging redirection of the institution's population and programs. There is no room for overlap or redundancy in the new and revised course offerings. The students must leave each course with measurable knowledge that is directly applicable to the career or graduate degree they pursue upon graduation. While it can be argued that this has always been the case, many of today's students are pursuing specific programs that give them a direct path to a position in an employment-ready field, and they are demanding means to acquire the skills and education without incurring excessive student loan debt.

At UW-SP, the processes for stakeholder collaboration and new content development are being identified and started. Meetings and discussions will be ongoing with appropriate science faculty at UW-SP, UW-Marshfield science and medical faculty (supporting the BAAS degree), the UW-SP Nursing Program Director (supporting the BS in nursing degree), the MSTC Associate Dean for the Service and Health Division (also supporting the BS in nursing degree), MCW representatives, as well as professionals in various medical settings. A campus visit by representatives from MCW enabled UW-SP faculty to discuss how they can support the new "Community Education Program" and revise or create new courses based on the needs of the program. This visit provided in-person contact with MCW stakeholders and formalized the affiliation between the program and UW-SP course offerings, including the new PW course. The UW-SP invitation to faculty to meet the MCW representatives focused on those teaching courses specific to the sciences, health and wellness, social work, and medical ethics. Professional writing intersects all of the content areas listed, since scientific and technical communication is an integrated skill in the field of medicine. Once the preliminary course goals are identified and defined by the appropriate parties, the new PW course plans can be solidified with a thorough analysis of the students' needs, based on the skills they will use in medical jobs, after they graduate. Informal job task analyses (JTAs) will be conducted with the medical workforce to identify areas that will benefit from targeted instruction. Through observation, JTAs focus on the individual tasks required by someone performing regular duties on the job. JTAs identify the necessary skills and knowledge required to successfully complete each task. These will be done in collaboration with current employees in the medical settings that are seeking graduates from the new programs. Employees will assist with the JTAs and identify competencies that new-hires will need when they start work in the medical field. Assignments for the new PW course will be created based in part on the JTA outcomes and identified competencies related to communicating and writing in medical settings. These competencies will be valuable resources for content development, implementation, assessment, and revision.

Collaborations aimed at determining specific content and the most useful approach to teaching it will be ongoing with UW-Marshfield science and medical faculty who are directly involved with the BAAS degree, as well as the UW-SP Nursing Program Director and the MSTC Associate Dean, regarding the BS in nursing. Collaborations with MCW Community Advisory Boards will provide guidance and oversight as the new MCW program advances. Additional research of successful programs in other colleges and universities will also be conducted to learn what works, and what to avoid, as they move through their own growth challenges. Wherever possible, technologies will be used to minimize interference in the daily flow of schedules and work commitments. This will likely include conference calls, collaborative software for professionals, and email. For the course content to be relevant and accurate, collaboration will be ongoing with colleagues in the new programs that are in development throughout the UW system. Working together will help guard against overlap and redundancy among courses offered by different campuses. UW System communication and advertising will be a key factor in informing students of the new courses offered by different campuses. Regarding the new PW course, the goal is that it will be embraced and added to the existing curriculum, and there may also be an opportunity to revise elements of the existing course content in UW-SP PW courses as well.

There are obvious tradeoffs that must occur when eliminating existing course content and replacing it with something new. Opportunity costs exist due to the omission of content to allow for the inclusion of new material and assignments focused strictly on field-specific educational needs. Working directly with the faculty who share the responsibility of the new offerings will help ensure the success of each revised course. Pilots of specific course content will help to define the best positioning of the material within the existing PW courses. Ultimately, completely new courses, beyond the single contribution currently being planned, will be proposed, depending on program need, instructional staff, and scheduling. DiRenzo (2010) adds credibility to this projection that, PW programs "can provide colleges with the rhetorical tools—the practical and intellectual tools and techniques—to forge a new humanism suitable for the perils and promises of a new century" (p. 252). The local result will be new and revised PW courses specifically designed for the medical students who, upon graduation, will provide quality healthcare services made available throughout Wisconsin's communities.

Phase 3: Transformation

Becoming hurts.
—Kat Howard (n.d.)

In the Transformation phase, new processes and structures are integrated into the existing model (Newman, 2012, p. 68). Concrete, measurable changes are seen in the transformation phase of the organizational change process. In terms of the curricular changes ahead for UW-SP, transformation will be apparent once the new and revised courses are approved and students are successfully completing them. In order to have the new PW offering, the instructor and department must discuss when it will be offered and who will teach it. New curriculum goals and course plans are developed and merged into established plans. The new course might be offered once a year until the demand for the course is fully proven. Once it is determined that the course can be added to the existing course schedule, the required paperwork is submitted to the department. The request to approve a new offering requires numerous details, including the course title; number; description; when it will first be offered; expected enrollment; who will be teaching it; the rationale and need for it, including assessment evidence; and a host of other questions that must be answered in writing before sending it on to the appropriate department members and university committees.

The new PW course will be proposed as a 3-credit, 300-level course. The course description will include a reference to an XL component of the course, which will help bridge academic writing to health-related fields. The XL component requires further collaborative efforts with the community coordinators who are involved with the course. Time will be given in class to discuss issues that arise throughout the XL activities, including class time with the community coordinators, when their schedules permit. Once the discussions among all stakeholders are complete, specific content areas, assignment details, and textbooks will be determined for the new course.

Assessment of course objectives and measurable learning outcomes will be a required and ongoing activity within the new PW course offering. In an extensive study by Aimee Whiteside (2003), recent graduates from undergraduate technical communication programs were surveyed, and it was concluded that "further research is needed at local, state, and national levels to analyze technical communication undergraduate curricula along with responses from recent graduates of technical communication programs and managers of technical communication programs" (p. 1). In the future, a study much like Whiteside's will be required at UW-SP to "provide empirical evidence for educational institutions looking for data to justify changes to their technical communication programs, and [help] future technical communication practitioners determine the skills they need to more easily transition into business and industry" (p. 3). Assessment of the new PW course and future curricula at UW-SP will be an ongoing part of the growth process.

Collaboration among stakeholders is necessary, along with the tensions and tradeoffs certain to be experienced throughout the curricular planning process. Furthermore, the business of higher education requires a collaborative model that is absolutely necessary for academic institutions and departments to pursue, not only to be successful, but for pure survival, in order to ensure that the needs and requirements of today's college graduates are met before they enter the job market.

CONCLUSION

Hopefully, by sharing this information, it will encourage discussion and promote camaraderie and alliances with other institutions that are considering or undergoing change—change that will impact many disciplines across the campus, but most particularly, the curricula of technical communication programs. Easing colleagues' fears and navigating tensions, both expressed and unarticulated, take patience, a cooperative and resilient spirit, and adaptive behaviors required by successful pioneers of tomorrow's technical writing programs. As in all things, success and survival depend on the willingness to make necessary changes and to identify key stakeholders that can guide others confidently through the process of those changes, one step, or syllabus, at a time. The significant undertaking of path redirection requires the capability to modify the planned course of action, sometimes with swiftness, while navigating through the new landscape. New goals and creative, cooperative ways to achieve them are all part of the necessary journey. It is within this context that the pedagogical redirections of one institution were presented here—to serve not as a roadmap for all, but as the emerging path of one.

REFERENCES

Carlson, L. T. (2011). UWSP must evolve to meet changing environment, Summers says. *The Pointer*. Retrieved from https://www.uwsp.edu/pointeronline/Pages/articles/UWSP-Must-Evolve-to-Meet-Changing-Environment,-Summers-says.aspx

Chirac, J. (n.d.). *Brainy quote*. Retrieved from http://www.brainyquote.com/quotes/quotes/j/jacqueschi192957.html

DiRenzo, A. (2010). The third way: PTW and the liberal arts in the new knowledge society. In D. Franke, A. Reid, & A. DiRenzo (Eds.), *Design discourse: Composing and revising programs in professional and technical writing* (pp. 243–253). Fort Collins, CO: WAC Clearinghouse/ Parlor.

Emerson, R. W. (n.d.). Written a book? *Thinkexist.com.* Retrieved from http://thinkexist.com/quotation/not_in_his_goals_but_in_his_transitions_is_man/170475.html

Hall, D. J., & Derby, S. K. (2012, November 19). Gov. Scott Walker unveils agenda for Wisconsin during speech in California. Wisconsin State Journal. Retrieved from http://host.madison.com/news/local/govt-and-politics/gov-scott-walker-unveils-agenda-for-wisconsin-during-speech-in/article_a35a1378-31ed-11e2-bb6c-0019bb2963f4.html

Henze, B., Sharer, W., & Tovey, J. (2010). Disciplinary identities: Professional writing, rhetorical studies, and rethinking "English." In D. Franke, A. Reid, & A. DiRenzo (Eds.), *Design discourse: Composing and revising programs in professional and technical writing* (pp. 63–86). Fort Collins, CO: WAC Clearinghouse/Parlor.

Howard, K. (n.d.). Kat Howard. *Good Reads.* Retrieved from http://www.goodreads.com/author/show/4098207.Kat_Howard

Hyland, N., & Kranzow, J. (2012). Innovative conference curriculum: Maximizing learning and professionalism. *International Journal for the Scholarship of Teaching and Learning, 6*(2), 3.

Morey, A. I. (2004). Globalization and the emergence of for-profit higher education. *Higher Education, 48*, 131–150.

Newman, J. (2012). An organizational change management framework for sustainability. *Greener Management International, 57,* 65–75.

Rich, A. (n.d.). Written a book? *Thinkexist.com.* Retrieved from http://thinkexist.com/quotes/Adrienne_Rich/

Summers, G. (2012). *Our once and future public university: A call to action.* State of Academic Affairs address presented at University of Wisconsin-Stevens Point, Stevens Point, WI. Retrieved from http://www4.uwsp.edu/media/admin/provost/2012/soaaa2012.htm

University of Wisconsin-Stevens Point (UW-SP). (2010). *Welcome to Schmeeckle Reserve.* Retrieved from http://www.uwsp.edu/cnr-ap/schmeeckle/Pages/home.aspx

University of Wisconsin-Stevens Point (UW-SP). (2012a). Student learning outcomes. *Welcome to the English Department.* Retrieved from https://www.uwsp.edu/english/Pages/default.aspx

University of Wisconsin-Stevens Point (UW-SP). (2012b, May). *Healthy Communities initiative.* Retrieved from https://www.uwsp.edu/AcadAff/Pages/healthyCommunities.aspx

University of Wisconsin-Stevens Point (UW-SP). (2012c, June 25). *Medical College of Wisconsin makes selection.* Retrieved from http://www.uwsp.edu/urc/news/Pages/Medical-College-of-Wisconsin-makes-selection.aspx

University of Wisconsin-Stevens Point (UW-SP). (2013). *Writing and editing internship.* Retrieved from http://www.uwsp.edu/english/pages/studentResources/internship.aspx

University of Wisconsin-Stevens Point (UW-SP). (n.d.). *Partnership for thriving communities.* Retrieved from http://www.uwsp.edu/acadaff/Pages/thrivingCommunities.aspx

University of Wisconsin System. (2012). *Campuses.* Retrieved from http://www.wisconsin.edu/campuses/

Whiteside, A. L. (2003). The skills that technical communicators need: An investigation of technical communication graduates, managers, and curricula. *Journal of Technical Writing and Communication, 33*(4), 303–318.

CHAPTER 11

Using Situational Advantages Strategically to Address Challenges Faced in Creating a Bachelor of Science in Technical Communication in an Environment of Austerity

Corinne Renguette, Marjorie Rush Hovde, and Wanda L. Worley

In developing a new bachelor's degree in technical communication at Indiana University-Purdue University Indianapolis (IUPUI), we faced several challenges: some of them typical of conditions in other programs and some atypical due to new legislation mandated by the state of Indiana. As a service program in a unique setting, we found few published resources to guide us as we created this new bachelor's degree. Here, by sharing our experiences and the ways we strategically addressed the challenges, we hope to help others as they develop and redesign programs.

In this chapter, we will discuss how we addressed three specific challenges:

1. New state laws and requirements
    - bachelor's degrees capped at 120 credit hours
    - all degree programs include a general education common core of 30 credits in specified curricular areas to be completed ideally in the first 2 years

- all new degree programs have an articulation agreement with the state community college system (Indiana Commission for Higher Education)
- a new degree-proposal format focused on economic impacts (Indiana Commission for Higher Education)

2. Institutional expectations to build a new program, increase enrollment, and offer more online, hybrid, and nontraditional delivery methods in our program without increasing resources
3. A field that changes rapidly due, in part, to rapid changes in technology

We will also discuss two major advantages we had while developing our bachelor's degree, even in an environment of austerity: nonmonetary institutional support and our location in a large urban area. Although we recognize that our situation is unique, we believe our experiences can help other programs that may feel limited by the challenges they face to develop strategies for program development in the face of austerity.

BACKGROUND

With more than 30,000 students, Indiana University-Purdue University Indianapolis (IUPUI) is the third-largest university campus in Indiana and the second-largest campus in the Indiana University (IU) statewide system. A variety of forces have shaped the campus's current standing and culture and ultimately the conditions in which we created the new degree.

For many years, Indiana's two land-grant universities, Indiana University (with its main campus in Bloomington, Indiana) and Purdue University (with its main campus in West Lafayette, Indiana), offered courses at two branch campuses in Indianapolis, the state capitol. In 1969, the two campuses merged into one institution, IUPUI. Since Indiana University (IU) is the managing partner of IUPUI, the campus is considered a part of the Indiana University system, with all administrative and budget functions handled by IU. The IUPUI campus does not award its own degrees. In general, schools associated with IU award IU degrees; schools associated with Purdue award Purdue degrees. Of 21 schools and academic units that grant degrees in over 250 programs on campus, the Purdue University name is on only two of these schools: the School of Engineering and Technology and the School of Science. These two schools, however, are two of the largest schools on the IUPUI campus.

Our Technical Communication (TCM) program resides within the School of Engineering and Technology (E&T). (For more information, see www.engr.iupui.edu/tcm/.) ABET, the accreditation body for engineering and technology programs, includes the ability to communicate effectively as one of its student learning outcomes. So as part of the accreditation requirement, our program has provided professional technical communication service courses to engineering and technology majors for many decades. Over the years, the program has stood alone, with the program director reporting directly to the dean of the school, but has also been a part of several different departments. In July 2012, the School reorganized the technology departments. As a result, our program now resides within the Department of Technology Leadership and Communication.

Throughout its history, the TCM program has enjoyed strong support from Engineering and Technology colleagues who recognize the importance of effective technical communication in professional practice and from Engineering and Technology ABET accredited programs, which must provide evidence of good communication skills as a student learning outcome.

The TCM program has consistently functioned with no more than three full-time faculty members. With no graduate program from which to draw instructors, we have always relied heavily on part-time faculty (10 to 15 per semester), many of whom are working professionals in technical communication or related fields and many of whom have taught in the program for several years.

Over a decade ago, the TCM program expanded its mission of offering only service courses to engineering and technology majors and other interested students to offering an undergraduate certificate. We had the support and cooperation of several departments across campus and of the local chapter of the Society for Technical Communication. This certificate program has served a small number of students over the years, preparing them for entry-level positions as technical communicators and/or professionals proficient in technical communication.

In 2010, we significantly revised the certificate to better meet the needs of our stakeholders. We called on several members of our TCM Industrial Advisory Board to participate in this revision. During the process, a subcommittee of the board collaboratively created a matrix of the knowledge, skills, and professional qualities that technical communicators should possess. In creating this matrix, the subcommittee recognized that students would need several more years to develop these complex qualities than an undergraduate certificate program would allow. As a result, the subcommittee indicated on the matrix what knowledge, skills, and professional qualities students should develop at the certificate level, at a bachelor's degree level, and at a master's degree level.

As the certificate revision progressed, members of the Advisory Board encouraged us to develop a bachelor's degree in technical communication because they believed that such a degree was needed to prepare students adequately for the challenges of the profession. In the Central Indiana region, many technical communicators work for a variety of corporations, and so the Advisory Board members recognized a need for well-educated technical communicators. Although the Great Recession occurred in 2008, Indiana and several other Midwestern states felt its effects less strongly than did other areas of the country. In addition, several biomedical companies in the region that employed technical communicators were not hit as hard as some other regional companies, so technical communication career prospects remained relatively strong in the region.

While creating the degree program, however, we faced several challenges that added many hours to our workload and months to our timeline. Fortunately, we also enjoyed several advantages that we were able to use in order to address these challenges. Although each institution faces different situational factors, we believe that awareness of our process of addressing challenges and identifying and using available resources can be beneficial to other institutions that want to create or revise a degree program in technical communication.

In the next sections, we discuss these challenges and how we addressed them, drawing on our contextual resources. These challenges include (a) new state laws and requirements; (b) institutional appeals to offer more online, hybrid, and nontraditional delivery methods while increasing enrollment without increasing resources; and (c) a field that changes rapidly.

STATE FINANCIALLY DRIVEN CHALLENGES TO THE PROCESS OF CREATING A NEW BS DEGREE PROGRAM IN TECHNICAL COMMUNICATION

The first challenge was directly or indirectly financially motivated and required us to be resourceful in creating the Technical Communication BS degree proposal.

New 120-Credit Degree State Mandate

As we were developing the new technical communication degree proposal between 2010 and 2012, several financially motivated forces at the state level constrained our work. First, the state legislature passed a law, House Enrolled Act 1220, effective July 1, 2012, which required Indiana public 4-year colleges to limit bachelor's degree programs to 120 credit hours (State of Indiana, 2012). A program could provide an exception proposal to require more than 120 credits; however, the argument needed to be compelling and based primarily on requirements of external accrediting agencies. The reasoning behind this law was financially based—legislators perceived that students were taking too long to graduate, incurring too much debt, and postponing earning capabilities. They reasoned that if bachelor's degrees typically required only 120 credit hours, these problems would be lessened. This new state mandate meant that we needed to design a curriculum that could prepare students to become proficient technical communicators without surpassing 120 credit hours.

Although an early draft of our proposal outlined a program that required 122 credit hours, we decided to lower the credit hours to 120 to increase our chances of getting our proposal approved. Fortunately (or unfortunately) for us, no accrediting body guides the number or nature of the courses that technical communication degrees must include, so we did not have to appeal for an exception from this rule. Also, fortunately, we had designed the degree program to be flexible, so eliminating one course did not greatly diminish the potential quality of the preparation of the students.

By itself, limiting our program to 120 credit hours was not a huge roadblock; however, the state also mandated a new 30-credit common core of classes that would be accepted at all public colleges and universities in Indiana. The two together created a bigger challenge.

New General Education Common Core State Mandate

In response to another new Indiana law (Senate Enrolled Act 182), public higher education institutions in Indiana developed the Statewide Transfer General Education Core (STGEC) (ICHE, 2012), effective fall 2013. The STGEC allows a student who satisfactorily completes a 30-credit approved general education common core at any public Indiana higher education institution to transfer those 30 credits to any other public Indiana higher education institution to satisfy the general education core requirement (ICHE, 2012). The motivation behind this law was the perception that students were losing money when they transferred because their new institutions required them to take courses similar to ones they had taken and wished to transfer from their former institutions.

Fortunately, the law requires courses from competency areas, not one-to-one transfers of exact courses. For instance, if a student takes a basic general education humanities course from an approved list at the first institution, the second institution is required to count that course toward the general education core humanities requirement. Furthermore, students are encouraged to complete all of the 30 core course credits in their first 2 years of full-time study so that transfers can be completed smoothly.

This law was not as challenging to implement in the new TCM bachelor's degree plan of study as we feared. Fortunately, we had designed our curriculum to be flexible and interdisciplinary, so we were able to meet this requirement with minimal effort. The most difficult part was figuring out which area of our degree each of the required general education common core courses would best match. Our BS degree includes four core areas of curriculum: 53 hours of technical communication and related communication courses; 34 hours of science, technology, engineering, and math (STEM) courses; 12 hours of organizational and cultural dynamics courses; and 21 hours of humanities, languages, and social science courses. Each of these four areas includes both required courses and electives. Some of the general education common core courses were already required courses in our degree. Those were easy.

The largest area that needed to be adjusted in our plan of study was the humanities, languages, and social sciences area (HLSS), which had 15 hours available in HLSS electives. We originally required at least half of those courses to be at the 300-level or above, but the common core requires 12 hours of courses in this area, and most of the courses on that list are 100- or 200-level courses. The 12 hours of common core include one arts and humanities course (3 credits), one social science course (3 credits), one cultural understanding course (3 credits), and one from either arts, humanities, or social sciences (3 credits) to meet the common core requirements. That left only 3 hours for an elective course, so we specified that elective needed to be a 300-level or above.

The hardest part about incorporating these changes was making them all very clear on our plan-of-study form. After multiple revisions, we decided to highlight all the general education common core courses in yellow on the plan of study so it would be clear to students and advisors and easy to make sure the core requirements were met within the first 2 years of the degree. The feedback we received from advisors on this newly revised plan of study said that it was one of the easiest plans to follow because of this format.

In addition to reworking the degree requirements to include the general education common core requirements, we also needed to prepare an articulation agreement with Indiana's state community college as required by the Indiana Commission for Higher Education.

Required Degree Articulation Agreement with the State Community College

A third financially based constraint that influenced the approval process for our new BS degree is that the Indiana Commission for Higher Education requires all new degree programs to create articulation agreements with the state community college system. These articulation agreements are required so that students who earn associate's degrees at the community college can transfer seamlessly to an in-state, public 4-year institution to earn a bachelor's degree. Again, this practice has a financial motivation; the goal is for students to be able to transfer the maximum number of courses from the community college system to the public, in-state 4-year institutions so that they do not incur unnecessary tuition expenses.

The process of drafting and revising this articulation agreement between IUPUI and the state community college system added many months to the proposal creation and approval process. It was necessary to make several revisions of the draft and obtain signatures from busy upper-level administrators at both institutions. Typical of large higher education institutions, we had to work through a specific IUPUI campus office whose staff members are responsible for interactions and articulations between the two institutions. Specifically, we had to gain approval from the provost of the community college system, who was traveling internationally at the time. We would submit requests and then experience long delays in receiving responses. After approximately 5 months, we asked if the provost of the community college system had a deputy who could oversee the process, and that individual entirely rewrote our description of how the professional writing degree from one community college campus would articulate to our new technical communication bachelor's degree. At that point, we signed off on the agreement even though it was not written exactly as we wanted. While we believed that very few people from that one campus would continue their degrees at IUPUI, the financially motivated rule meant that we had to spend a great deal of time and energy to complete this piece of the approval process.

Despite long delays in working through this articulation agreement with the state community college system, we want to stress that our experience was not necessarily typical. Also, we strongly support articulation agreements between the state community college system and 4-year state higher education institutions. These agreements are vitally important to the success of thousands of students.

Now we turn to the fourth major challenge we faced in this area: a newly mandated proposal format that focused on economic impact.

State-Mandated Degree Proposal Format That Focuses on Economic Impacts

The Indiana Commission for Higher Education requires programs that propose new degrees to address financially related questions of the employability of graduates, the numbers of anticipated students, and the prospects that the new degree program will be self-supporting. All of these demands required research that could be used to support the case for the new degree. Some of them, such as how many majors we might anticipate, were almost impossible to answer with any degree of reliability. In writing the proposal, we were also keenly aware that we would have to propose that the new degree program be offered with mostly existing resources and that the Commission was unlikely to approve proposals that asked for additional start-up or ongoing funding. Because we already had a strong service program, we anticipated that we could offer the degree with existing resources, at least for the first year or two.

In addition, the Indiana Commission for Higher Education proposal format asks whether similar programs are offered on other state campuses, especially those in the IU or Purdue branch systems. The intent is to avoid threatening the enrollments of an existing degree program and to avoid duplication of existing degree programs. This financially related challenge required us to take considerable time to solicit multiple letters of support from other programs on our campus and throughout the state. Throughout the multilevel approval process, we frequently received suggestions to solicit even more letters of support from interested chairs and deans in order to enhance our chances of approval for the new degree program. Requesting these letters of support added greatly to the time required to create the proposal, although the support letters did make the proposal much stronger.

In the last weeks, just before the proposal was to be submitted we were told the Indiana Commission for Higher Education had just approved a new format that had to be used for all new program proposals. Our proposal had to be completely reformatted. Fortunately, we had help with the reformatting, a process that added only a few weeks to our timeline.

Once the proposal process was finally completed, the Indiana Commission for Higher Education approved our proposal very quickly with no suggested changes. While the entire process took much longer than we anticipated, we were excited to move forward. Of course, the challenges did not end there.

We next turned our attention to building our program.

INSTITUTIONAL EXPECTATIONS TO BUILD A NEW PROGRAM

In this section, we discuss meeting the institutional expectation that we build our program by increasing enrollment and offering more courses using non-traditional delivery methods without increasing resources.

Limited Resources

Because of budget constraints, the TCM program attempts to fulfill its mission with a small number of full-time faculty members supplemented by part-time instructors. The TCM program has never had more than three full-time faculty members, and we have no graduate teaching assistants because we do not yet have a graduate program. We are fortunate, however, to have a group of dedicated part-time faculty members, many of whom have been with the program for 10 or more years. While we appreciate their dedication to teaching in our program for very little pay per section, administrative and curricular development duties necessarily fall to the three full-time faculty members in addition to their ordinary duties in, among others, (a) overseeing the offering of service courses to other majors, (b) administering the TCM Certificate, (c) running a Technical Writing Center, and (d) participating in the assessment of students' communication skills for purposes of ABET accreditation.

This constraint of limited numbers of full-time faculty members challenged us as we developed the new degree and continues to present ongoing challenges as we develop new courses and revise old ones to meet the needs of the TCM BS students. We are unable to get additional help until our BS degree is large enough to justify an additional full-time faculty member.

Changes in Instructional Delivery Methods

With technology advancing exponentially, online education has continued to grow and change with the students. Students interested in online education started out as adults needing an alternative to sitting in a college classroom because of work and family responsibilities; now those interested include students of all ages, including high school students, who find online education appealing for a host of reasons from convenience to the appeal of mobility.

Motivating the move toward online and hybrid teaching and learning in our state are Indiana legislators who are moving more and more to performance-based financing, which means the state financially rewards state higher education institutions for the number of student graduated in 4 years, as well as a number of other performance-based outcomes. According to Kelderman (2011),

> The state of Indiana has moved aggressively toward performance-based financing in recent years. Five percent, or about $61 million, of the state's higher-education appropriation is based on a variety of performance measures, including credit-hour completion, the number of low-income students who graduate from an institution, and the number who earn degrees in science, technology, engineering, or mathematics, the so-called STEM fields. (pp. A12–A13)

IUPUI traditionally offered face-to-face night and weekend classes to accommodate part-time students who are employed during the day, but in recent years, administrators have encouraged faculty members to offer online and/or hybrid courses and to experiment with new nontraditional schedules for face-to-face courses to attract a wider variety of students. Traditional universities are facing stiff competition from other institutions that offer completely online courses, certificates, and degrees.

In 2008, about half of approximately 55 sections of our service courses were offered online. In 2013, we were still offering approximately half of the sections as fully online classes, with 37 out of 89 sections fully online. In addition 17 of the 89 sections were hybrids in which students met face-to-face only as needed and conducted the remainder of their work online. Combining the online and hybrid sections, well over half of the sections (54 of the 89) were delivered in an alternative delivery format.

Administrators often think professors can teach more students online than in a face-to-face classroom, feeding into the performance-based demands of the legislators. Teachers, however, know this is not true, as online courses that are taught well often require much more time and energy investment than their face-to-face counterparts. Writing classes are also more demanding in an online environment because of the amount of grading that is not able to be automated. Our school gives no special compensation or allowance for teaching online sections. Neither does it give special recognition for online teaching in the promotion and tenure process.

Classroom space is at capacity at IUPUI and, as we have indicated, funding is limited. Largely for financial reasons of increasing enrollment without increasing resources, the IU system has launched several initiatives encouraging faculty members to offer courses using nontraditional delivery methods. For example, in 2012, IU announced *IU Online*, "a major strategic investment in online education that will accelerate the development and delivery of quality online courses and programs at IU's campuses statewide, address Indiana's economic and professional development needs, and extend the university's global reach" (Indiana University, 2012). IU will spend $8 million over 3 years on this initiative. Again, one of the motivators is to meet the performance-based outcomes demanded by our legislators: "IU Online will also focus on reducing the time it takes students to complete degree programs by making certain online courses available throughout the year and exploring the development of 'massively online' undergraduate courses" (Indiana University, 2012). To prepare for this IU online initiative, IU established the Office of Online Education in spring 2011 "to provide leadership, coordination and quality assurance of online education across all IU campuses based on a recommendation of the IU Strategic Plan for Online Education" (Indiana University, 2012).

Our goal is to offer the majority, if not all, of the BS degree courses in both an online and face-to-face delivery format. Administration at the campus and school levels encourages us to offer as many courses as is feasible online. Students who have expressed interest in the degree since its approval in November 2012 have consistently expressed a preference for online delivery. A recent development that may negatively affect our online course enrollment is that international students (about 10% of our Engineering & Technology students) who are sponsored by their home governments are restricted to the number of online courses they can take.

If our new degree program is to be financially sustainable, we will need to accommodate the increasing demand for online courses while still providing the highest quality education possible. As many who have developed and offered online courses can attest, the process of developing online courses takes a tremendous amount of time if the courses are pedagogically sound, which costs money if faculty are compensated for developing the classes. The demands of offering the online course also usually require more, not less, attention per student from the instructors than face-to-face formats require (Worley & Tesdell, 2009). The challenges of offering courses in alternative formats have direct and indirect effects on the financial dynamics of sustaining the quality of the degree.

So the challenge of limited resources continues to constrain our progress. In addition to the limited number of faculty members, we are also constrained by the rapidly changing field of technical communication.

A FIELD THAT CHANGES RAPIDLY

In addition to financially motivated state and institutional constraints, in developing our new BS program, we had to consider the rapidly changing nature of technology and technical communication. A Bachelor of Science in Technical Communication must anticipate these changes so we can teach content that will be relevant when students graduate.

Changes in Technology

In addition to preparing students to understand concepts central to technical communication and to be critical thinkers, "those in academia must be aware of the skills and abilities new writers will need to succeed in the workplace of today and in the future" (Rauch, Morrison, & Goetz, 2010). Some of the recent technological advances that are relatively new to the field of technical communication (or changing rapidly in the field) include the development of and increasing use of social media, online customer support, wikis, blogs, videos, and more. In addition, using these new technologies can help technical communicators better meet accessibility and global communication needs.

With these changing technologies also comes a need for the technical communication faculty to continue updating curriculum and pedagogy (Savage, 2013) as well as their own knowledge base. One major change in the use of technology is that many types of communication have gone from print to electronic, including documentation. So where once we could teach students how to create a printed, document-based manual in Word, now we must consider that many manuals are created as online PDFs with XML and online help documentation (Rauch et al., 2010). Students (and faculty) need to be aware of and know how to learn new types of technology, so we need to focus pedagogical efforts on teaching not only how to learn about and use new technologies, but also how to think critically about and evaluate new technologies. We designed a new sophomore-level course, Tools for Technical Communication, to introduce students to relevant concepts and teach them critical thinking skills in relation to technology, strategies for learning new technologies, and concepts and skills that they will apply in upper-level courses and in their lives after graduation.

Keeping current in technology software also presents financial challenges to our program and school. In response to this challenge, several departments in our school have begun requiring majors to provide their own laptop computers and purchase their own software to meet the specific needs of the programs. The main benefit to this is financial: Classes do not need to be held in a computer lab if students have their own laptops, reducing the university's cost for adding and maintaining computer labs. Fortunately, IU has an arrangement (called IUware) with several major software companies that allows students, faculty, and staff to obtain common software applications at low or no cost. As we designed our degree program, we decided to require majors to provide their own laptop computers. They can then download or purchase (at a significantly reduced price) any necessary software through the IUware program. Although there are many software applications available to technical communicators, for financial reasons, we elected to use in our courses only those applications accessible to our students and faculty members through IUware.

In designing our degree program, we made decisions about how to ensure that students would have access to current technology in order to meet educational goals and to prepare for changes in technical communication careers.

Changes in Technical Communication Careers

In addition to changes in technology, the field of technical communication has undergone changes itself. Early in the field, the majority of students were trained as technical practitioners who needed to learn how to become communicators; now the field has become a profession of its own, not just supporting the work of people in other disciplines (Savage, 2013). Despite the field becoming more established, many challenges remain. According to survey respondents who had completed graduate coursework in technical communication, academics value master's degrees, but often employers do not, and there is a "lack of 'respect' for technical communicators' work" in the field (Bloch, 2012). Respondents also noted how important it is for technical communication programs to "offer plenty of hands-on, real-world projects" and "incorporate appropriate technologies" (Bloch, 2012, p. 7). Communication between authors and customers has also changed, in addition to "customer troubleshooting and information sharing . . . and faster product development and communication" (Rauch et al., 2010, p. 297). Furthermore, as economic conditions decline, many companies reduce numbers of technical communicators as a cost-saving measure. These demands add the challenge of teaching students in a degree program how to work with clients and manage projects efficiently, as well as shaping arguments about the value of effective technical communication.

To be successful in a technical communication career, students need to learn multiple skillsets. The field "requires training specifically in technical communication but also in visual communication, document/web/multimedia design, project management, content management, usability and user-centered design, and activity theory" (Savage, 2013, p. 11). Career paths in technical communication include positions such as Corporate Trainer, Change Management Consultant, Software QA Engineer, IT Specialist, Grant Writer, Information Developer, and more (Bloch, 2012). Technical communicators can work in fields such as health care, law, academia, software, hardware, pharmacy, science, and more (Bloch, 2012). This wide variety of content requires students to learn how to be adaptable, especially in rapidly changing economic circumstances.

In order to help students learn about the field and also be prepared for technological changes that they might face in their careers, degree programs need to hire faculty who may be current practitioners and collaborate with them so they can "adapt to change and incorporate new tools, media, and methods of development and delivery," the same way that technical communicators do (Rauch et al., 2010, p. 306). Programs need to offer internships and practical experience communicating in realistic settings and to teach students how to make connections with others in the field so that they can address future uncertainties.

As with technological changes, these changes in the field also have financial implications for the development and administration of technical communication degree programs. In designing our degree program, we needed to factor in the eventual need for more full-time faculty members with areas of specialization relevant to the new degree program, an extension beyond the skillset faculty members need in order to oversee and teach the TCM service courses. In addition, we needed to expand our existing network of practitioners in order to provide internship opportunities to meet the requirements of the new degree. One financial implication of internships is that our school's Office of Career Services is encouraging us to ensure that all internships are paid in order to avoid several common problems with unpaid internships. Therefore, we need to make connections with local organizations that can afford to pay at least a modest stipend for the work of interns we place with them.

All of these financial constraints—the state and institutional restrictions along with changes in technology and the field—require the strategic use of any available contextual advantages.

CONTEXTUAL RESOURCES USED IN DEVELOPING OUR BACHELOR OF SCIENCE IN TECHNICAL COMMUNICATION

Although the TCM program faced considerable financially related challenges in designing and developing its new bachelor's degree, we were also able to use several contextual resources and advantages to strategically address those challenges. Finding and using these resources was central to overcoming the challenges discussed in the previous sections. Specifically, we were able to benefit from institutional support and a campus located in an urban environment, as discussed below.

Institutional Support

The most immediate contextual influence on the development of a new degree program perhaps lies in the immediate department or school in which the program will be housed. Academic administrators are known for wanting new initiatives to use existing resources. We surmised that if we could design a program that used existing courses and instructors, our chances for approval would be stronger. Because of our situation, we were able to use resources such as administrative support, our home within a school of engineering and technology, and our existing strong service program and undergraduate certificate.

Administrative Support

The proposal for the new degree was developed during the last year of a retiring engineering and technology dean and the first years of a newly hired one. Both administrators provided us with financial and moral support for the proposed degree program, support that was extremely important in persevering through the challenges of developing a new degree program. During the process, we received funding for a part-time program assistant who was able to interview local technical communicators and complete background research about regional career opportunities, competing degree programs, and the nature of students who we were likely to attract. One of the deans mentored us through the process of creating the proposal, making sure we looked at every angle possible. His expert guidance helped ensure the eventual success of our proposal.

Home in an Engineering and Technology School

Our location in the School of Engineering and Technology (E&T) is another advantage we were able to use as we created a degree program that was interdisciplinary in nature that includes a large number of STEM (science, technology, engineering, and math) courses, elements of the engineering design process to strengthen the courses, and valuable access to industry partners. Because we are located in the context of the School of Engineering & Technology, it made sense to create the degree as a Bachelor of Science rather than a Bachelor of Arts. We were well supported in our decision to require a substantial portion of STEM courses, a decision that many have told us will help our students be more likely to gain respect in the workplace. Another advantage to being in the School of E&T is that we can incorporate elements of the engineering design process into our instruction and have it be well understood. A third advantage is an active Dean's Industrial Advisory Council, which provides us with networking opportunities as well as internship opportunities for students. In our degree design process and in our future development of the degree program, we plan to utilize these and other resources that our School offers.

In addition, being located on a large campus with a diversity of programs allowed us to address the challenge of a limited number of full-time faculty members in the program by creating the degree requirements to take advantage of existing relevant courses on campus. To supplement a strong core of technical communication courses, we were also able to require or encourage students to take courses from related departments that would enhance their abilities to communicate in organizational settings. In addition, we required a given number of courses in a technical and/or scientific field, thus enriching the students' backgrounds in ways that will be useful for professional practice. In order to equip students to deal with organizational and cultural variables in technical communication practice, for our third core area we require students to take courses in organizational leadership and supervision and other related fields. We also use our campus's extensive offerings in languages and the liberal arts to round out the fourth core area with broad-ranging knowledge that will enable them to face multiple challenges throughout their careers.

Our university also has several valuable resources to help us stay current with the changes in technology. In addition to being able to access the large number of software applications through IUware, IU's University Information Technology Services (UITS) offers students and faculty free access to lynda.com, which provides hundreds of training videos teaching people how to use a broad variety of software programs. UITS also provides face-to-face classroom training workshops, Web-based training, and access to self-paced online training materials. Our on-campus Center for Teaching and Learning (CTL) is fully staffed to help with curriculum development and pedagogical concerns. We were fortunate to be awarded a Curriculum Enhancement Grant through the CTL that helped us develop several of the foundational courses for the major to be offered online. We also used resources from UITS and CTL to aid us in creating several new courses for the major.

To address the challenge of teaching students how to work with clients and manage projects efficiently, we are able to use our university's Solution Center, which acts as a liaison between business and academia. The Solution Center matches an organization's needs with departments or programs on campus that can help meet those needs. In collaboration with the Solution Center, we offer students a variety of experiential class projects and internship opportunities to help them gain much-needed authentic experience.

Using these school and campus resources has helped us overcome some of the financial and personnel limitations inherent in our setting. While not all universities necessarily have all of these resources, they will have their own unique resources and opportunities. Investigating those resources available on one's campus is critical to a successful proposal. Administrators are highly suspect, and rightfully so, if the proposal suggests duplicating existing school or campus resources.

Strong Service Program and Existing TCM Certificate

Because of our strong service program with approximately 87 sections of 14 fully developed oral and written communication courses in topics ranging from general workplace technical communication to technical communication in engineering, technology, and even healthcare professions, we were able to use those courses as a basis for the requirements for the new bachelor's degree program.

In addition, our dedicated group of part-time instructors meant that we had many experienced people available to teach these courses and the new ones that we develop for majors. Having administrative support, being located in a school of engineering and technology, and having a strong service program and existing technical communication certificate were huge advantages for us in developing our bachelor's program. Without these resources, developing the degree would have been much more difficult and taken even longer than it did.

Location in an Urban Setting

In addition to staying current with new technologies, we found valuable regional resources to help us keep up with the changes in our field. Because of our setting in the metropolitan downtown Indianapolis area, we have access to local talent who are current in the field, who serve on our Industrial Advisory Board, and/or teach classes as adjunct instructors. Early in the process of creating the degree, we were able to interview several technical communicators or their supervisors in order to understand current needs in the field. Later in the process of developing the degree program, we invited members of our Industrial Advisory Board to review drafts of documents that described the knowledge, skills, and personal qualities that graduates of the program would need to develop. In addition, while we were developing or revising new courses for the degree program, we invited them to review the objectives and content of those courses. Their feedback helped us greatly to refine the direction of the curriculum.

This urban setting gives us abundant resources for future internship possibilities and eventual job opportunities for students upon graduation. Leveraging the resources available within the local community can greatly strengthen any program.

USING RESOURCES STRATEGICALLY TO ADDRESS CHALLENGES

In an environment of austerity, developing our Bachelor of Science in Technical Communication had its challenges; however, we also had access to resources that we were able to draw upon to address many of the challenges. Table 1 offers a brief summary of how we used resources strategically to meet the challenges.

Table 1. Brief Summary of Challenges Addressed Strategically

Challenges	Methods/Resources Used to Address Challenges
120 credit hours requirement	Focused the degree program
General Education Core Curriculum	Modified an already-developed flexible plan of study for the major
Articulation agreements with the state community college	Collaborated with the IUPUI "Passport" office responsible for interactions with the state community college
A state-mandated degree proposal format that focuses on economic impacts	Received administrative support to pay a student researcher to address issues of employment prospects, prospective students
Rapidly changing technology	Used available software applications through IUware and used other campus resources such as University Information Technology Services (UITS) training opportunities and access to lynda.com videos
Rapidly changing careers in technical communication	Consulted with our network of regional professionals interested in technical communication
Institutional demands for online, hybrid, and non-traditional delivery of courses	Used IUPUI's Center for Teaching and Learning staff and their Curriculum Enhancement Grant to develop and revise courses
Limited financial resources	Utilized part-time faculty (practitioners in the field) to teach courses Included existing courses in TCM, in our school, and on campus to create an interdisciplinary, flexible course of study Consulted with regional professionals to guide curricular development

Although we had to work within the challenges and sometimes make compromises, we also found that by using resources strategically, we were able to create a new major that offered students strong preparation in technical communication.

CONCLUSION

Any academic context presents financially related challenges that may discourage the development or revision of degrees and curricula. In developing our new program, we faced typical challenges as well as the atypical ones such as state-mandated changes to programs that were unexpected and last minute. However, we used available resources strategically to overcome those challenges. Technical communication is an important field with far-reaching effects on people's use of technology, so finding resources to keep curricula up to date and as strong as possible is a necessary goal.

We recognize that each situation is unique, so the challenges and resources we have described in this chapter will differ from other academic contexts. Nevertheless, our discussion can provide ideas for faculty in other institutions as they analyze their situations in order to find creative ways to address contextual challenges. The task is not easy, but it is doable. In this process, we balanced our educational goals by working within the constraints of new financially driven challenges from the state, institutional challenges, and challenges of being in a rapidly changing field. However, we were able to use existing resources creatively to address those challenges and provide what promises to be a strong degree program for our new majors.

ACKNOWLEDGMENTS

We would like to thank these people without whose support and help we could not have developed our Bachelors of Science in Technical Communication: David Russomanno, Dean, Purdue School of Engineering and Technology; Oner Yurtseven, Former Dean, Purdue School of Engineering and Technology; Stephen Hundley, Chair, Department of Technology Leadership and Communication; Candiss B. Vibbert, Associate Provost, Purdue University, West Lafayette; Ellen Harley, TCM Assistant & Graduate Student; Allieson Serd, TCM Intern; and all those who wrote letters of support for the new degree.

REFERENCES

Bloch, J. (2012). Envisioning career paths in technical communication: A survey of participants in a technical communication graduate program. *Professional Communication Conference (IPCC), 2012 IEEE International*, pp. 1–8.

Indiana Commission on Higher Education (ICHE). (2012). Statewide Transfer Gen Ed Core. *IN.gov*. Retrieved from http://www.in.gov/che/2728.htm

Indiana Commission on Higher Education (ICHE). (n.d.). Home. *IN.gov*. Retrieved from http://www.in.gov/che/

Indiana University. (2012, September 5). *Indiana University announces IU Online, a major new online education initiative*. Retrieved from http://newsinfo.iu.edu/news-archive/23061.html

Kelderman, E. (2011). Online public university plans to turn Indiana dropouts into graduates. *Chronicle of Higher Education, 57*(28), A12–A13.

Online Colleges. (2010, January). *The history of online education*. Retrieved from http://www.onlinecolleges.net/2010/01/26/the-history-of-online-education/

Rauch, M., Morrison, C., & Goetz, A. (2010). Are we there yet? An examination of where we've been and where we're headed as technical communicators. *Professional Communication Conference (IPCC), 2010 IEEE International*, pp. 297–309.

Savage, G. (2013). Educating technical communication teachers: The origins, development, and present status of the course, "Teaching Technical Writing" at Illinois State University. *Communication and Language at Work. Technical Communication, 1*(2), 3–19.

State of Indiana. (2012). *House Enrolled Act No. 1220*. Retrieved from http://www.in.gov/legislative/bills/2012/HE/HE1220.1.html

Worley, W. L., & Tesdell, L. S. (2009). Instructor time and effort in online and face-to-face teaching: Lessons learned. *IEEE Transactions on Professional Communication, 52*(2), 138–151.

Contributors

AMANDA BEMER is an assistant professor at Southwestern Minnesota State University. Her primary research interests are rhetoric, visual rhetoric, and professional communication.

TIFFANY BOURELLE is an assistant professor at the University of New Mexico (UNM), where she teaches technical communication and first-year writing in both face-to-face and online media. Her research involves enhancing existing pedagogies for the online environment, including service-eLearning and multimodal composition. She currently serves as co-administrator for an online first-year writing program she helped develop at UNM called eComp (short for Electronic Composition). Bourelle's publications have appeared in such journals as *Computers and Composition, Kairos: A Journal of Rhetoric, Technology, and Pedagogy*, and *Technical Communication Quarterly*.

ANGELA CROW, an associate professor at James Madison University, has published in the area of assessment and big data, aging and literacy issues, and faculty development vis-à-vis technology and digital rhetoric training. She has recently focused her research on the rhetorical strategies that cyclists are choosing in online venues when advocating for rights to the road.

TIMOTHY D. GILES has been involved in technical communication for 30 years. He currently directs the Professional and Technical Writing area at Georgia Southern University. He is the author of *Motives for Metaphor in Scientific and Technical Communication* (2008), and his articles on metaphor, risk communication, and readability have appeared in *Journal of Technical Writing and Communication.* He received his Ph.D. in rhetoric, scientific, and technical communication from the University of Minnesota–Twin Cities, and M.A. in English, technical and professional writing from East Carolina University.

TERESA HENNING is an associate professor of English at Southwest Minnesota State University in Marshall, Minnesota. She directs the Professional Writing and Communication Undergraduate Major, is interim director of the University Writing Center, and is coordinator of the AAC&U/Minnesota

Collaborative Pilot Project. Her research interests include writing program administration as it relates to undergraduate writing majors and writing with technology. Henning has published articles or chapters in *Journal of Asynchronous Learning Networks, English Journal, Journal of Effective Teaching, Writing Lab Newsletter, Writing and the iGeneration, What We Are Becoming,* and *Peer Pressure, Peer Power.*

MARJORIE RUSH HOVDE is an associate professor of technical communication in the School of Engineering and Technology at Indiana University–Purdue University Indianapolis (IUPUI), where she coordinates the Technical Communication Certificate and has developed a variety of undergraduate and graduate courses in technical communication. She recently served as primary author for the proposal for the IUPUI Technical Communication Bachelor's degree discussed in this volume. Her research has appeared in Technical Communication Quarterly, Journal of Business and Technical Communication, and Programmatic Perspectives.

LYNN O. LUDWIG is an assistant professor of English at the University of Wisconsin–Stevens Point, where she teaches technical writing and business communication. Her approach to teaching is informed by over a decade of experience in managing and writing in corporate and government settings. Her current research and publication background is in the scholarship of teaching and learning, with ongoing participation in international collaborative writing projects. Passionate about usability testing, Ludwig teaches the subject from the viewpoint of her previous work as an IBM software education development manager and her volunteer work as a usability tester for IBM products.

BARRY MAID is a professor and founding head of the Technical Communication Program at Arizona State University. He led that program for ten years. Previously, he taught at the University of Arkansas at Little Rock, where, among other things, he helped in creation of the Department of Rhetoric and Writing. Along with numerous articles and chapters focusing on technology, outcomes assessment, information literacy, independent writing programs, and program administration, he is coauthor, with Duane Roen and Greg Glau, of The McGraw-Hill Guide: Writing for College, Writing for Life.

COLLEEN A. REILLY is a professor of English at the University of North Carolina Wilmington, where she teaches courses in professional and technical writing, including document design, technical editing, writing about science, writing and activism, and writing and technology. Her research interests relate to professional writing program development, digital composition pedagogies, digital research methods, power and legitimacy in professional and technical communication, and electronic open access publication.

DIRK REMLEY has taught technical writing and business writing courses at Kent State University for over twenty years. His research interests are in business writing and technical writing pedagogies, including Web-based pedagogies and multimodal composing. He has authored two books published in Baywood's

Technical Communication series, and his work has appeared in several journals and handbooks. He has also presented at national conferences.

CORINNE RENGUETTE is an assistant professor of technical communication and director of the Technical Communication Program in the School of Engineering and Technology at Indiana University–Purdue University Indianapolis (IUPUI). She received her Ph.D. in applied linguistics and has certificates in Technical Communication and Teaching English to Speakers of Other Languages. She has designed, revised, and taught a variety of courses in areas such as technical communication, English for academic and specific purposes (engineering, technology, legal, apprenticeship trades, intensive English), and applied linguistics. Her research focuses on health literacy and the use of technology in patient education materials.

ROCHELLE (SHELLEY) RODRIGO is an assistant professor of rhetoric and (new) media at Old Dominion University. She was a full-time faculty member for 9 years in English and film studies at Mesa Community College in Arizona. She researches how "newer" technologies better facilitate communicative interactions—more specifically, teaching and learning. As well as coauthoring two editions of *The Wadsworth Guide to Research,* she coedited *Rhetorically Rethinking Usability.* Rodrigo's work has also appeared in *Computers and Composition, Teaching English in the Two-Year College, EDUCAUSE Quarterly, Journal of Interactive Technology & Pedagogy,* and various edited collections.

JULIA ROMBERGER is coordinator of professional writing and director of the English Department Computer Lab at Old Dominion University. Her research interests include ecofeminist methodologies, rhetoric, technology, and digital writing.

JULIE STAGGERS is an associate professor of rhetoric and writing at the University of South Florida. She has published 10 articles and textbook cases and presented more than 30 papers on a variety of topics, including writing program administration, teaching in digital environments, technical communication, risk communication, and visual rhetoric. She has served as a reviewer for *IEEE Transactions on Professional Communication, Online Writing Instructor,* and *Xchanges: Journal of Student Research,* and as a conference reviewer for the Association of Teachers of Technical Writing.

JANICE R. WALKER is a professor of writing and linguistics at Georgia Southern University. She has published journal articles, book chapters, and books on computers and writing, research and documentation, tenure-and-promotion issues, and writing. She is founder and coordinator of the Graduate Research Network at the annual Computers and Writing Conference, and co-coordinator for the Georgia International Conference on Information Literacy hosted annually by Georgia Southern University. Her current research includes the LILAC Project (Learning Information Literacy Across the Curriculum), a study of students' online information-seeking behaviors.

WANDA L. WORLEY is an associate professor of technical communication in the School of Engineering and Technology at Indiana University–Purdue University Indianapolis (IUPUI). She served as director of the Technical Communication Program for 10 years before accepting the position of associate dean for academic affairs and undergraduate programs in 2013. She currently serves as a review board member for *Programmatic Perspectives* (a peer-reviewed electronic journal published by the Council for Programs in Technical & Scientific Communication, CPTSC) and as editor for the *International Journal on Recent Trends in Engineering and Technology*. Worley's research focuses on the scholarship of teaching and learning, online teaching and learning, social media, and efficacy of peer review.

MADELINE YONKER is an assistant professor of rhetoric and composition at York College of Pennsylvania. Broadly, her research and teaching interests include technorhetoric, professional writing and technical communication, digital culture, networks, narrativity, and composition-as-social.

MICHAEL J. ZERBE is an associate professor of English and coordinator of the Writing Program at York College of Pennsylvania, where he teaches first-year writing and many courses in the Professional Writing Major. With Dominic DelliCarpini, he has published two chapters on undergraduate writing majors, and he has also published several works on the rhetorics of science and medicine, including *Composition and the Rhetoric of Science: Engaging the Dominant Discourse* (Southern Illinois University Press, 2007). Zerbe was awarded a Fulbright to teach in Bulgaria in 2009 and a Health Communications Fellowship from the National Cancer Institute in 1990.

Index